W9-CQF-707

BIOLOGY 12
BC Curriculum
STUDY GUIDE
Fifth Edition

Copyright 2010 Prior Educational Resources Ltd.
Reprinted August 2011

ISBN 978-0-9811373-0-8

Author and Editor: Roger Prior

Cover Photograph: Donna Prior
Cover Design: Linda Mullin
    (Linda@MullinArt.com)

Illustrators: Roger Prior, Jacqui Lee
Artwork: Kellie Hipperson

Published by Prior Educational Resources Ltd.

Printed and bound in Canada by Friesens

TABLE OF CONTENTS

I – CURRICULUM

UNIT A - THE CHEMISTRY OF LIFE

Life, as we know it, requires water and is based on the chemistry of organic molecules. To be "**organic**" (as opposed to "inorganic"), a molecule must contain carbon. For example, vinegar (CH_3CH_2OH) is organic, whereas water (H_2O), something organisms depend on, is not. Cells and their organelles are made out of organic molecules commonly called **biochemicals**. There are four types of biochemicals: carbohydrates, lipids, proteins, and nucleic acids.

When chemical compounds form, atoms bond together to produce stable structures. To achieve this, the electron configurations around the various nuclei in the compound must also be stable. This stability is achieved by sharing electrons between the atoms. A **covalent bond** (such as the bond between carbon atoms in an organic molecule) forms when the atoms share the electrons reasonably equally. The alternative is called an **ionic bond** where one atom actually loses the use of one or more electrons to the gain of another atom. The atoms become oppositely charged and attract each other. Ions form when molecules that are ionically bonded together break apart, or **dissociate**. In between these two extremes, bonds are termed **polar covalent**. Molecules with polar covalent bonds have **dipoles**, regions with slight positive and negative natures (abbreviated δ^+ and δ^-, respectively).

SOME FUNDAMENTALS

WATER

Water is a polar, covalent, inorganic molecule. Because of the difference between the nuclei of oxygen and hydrogen atoms, the bonding electrons in a water molecule are not shared equally and the region around the oxygen nucleus has an overall negative nature to it relative to the rest of the molecule. This is called a **negative dipole**. For the opposite reason, the ends of the molecule where hydrogen is located have **positive dipoles**. Opposite dipoles attract, meaning water molecules will attract each other.

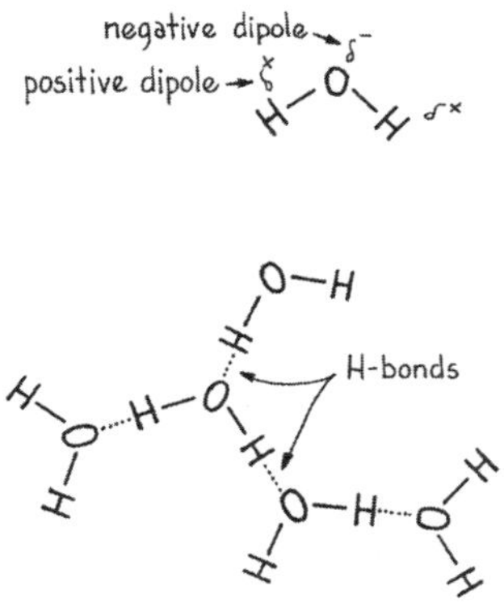

The weak forces that attract and hold molecules with opposite dipoles together are called **hydrogen bonds** (H-bonds). Water molecules are loosely held together by hydrogen bonds between the H of one molecule and the O of another. The stability of a hydrogen bond, like any bond, is affected by temperature. If water temperature decreases, the H-bonds holding the molecules together become more stable. If temperature increases, the kinetic energy increases and H-bonds can break.

Figure A-1. Polar Water Molecules. *A water molecule is bonded together by polar covalent bonds. This creates the positive and negative dipole regions marked above. Attraction between opposite dipoles results in H-bonding, which holds adjacent water molecules together and some interesting and important properties result.*

The **cohesion** among water molecules allows water to travel in tubes such as through roots and up to the leaves of plants and in the blood vessels of animals. Cohesion also accounts for surface tension, which is a phenomenon that provides some strength to the surface of water allowing surface dwelling creatures like water striders to scurry across the tops of puddles. Water has other interesting properties, several of which are directly significant to the human body.

Water is an excellent **solvent** for polar molecules. Molecules that are ionically bonded together will **dissociate** (break apart) in water and the water will transport the resulting ions. This is what happens in blood. Water also has a high **specific heat capacity**. Because of H-bonding, water either absorbs or withstands releasing a lot of energy before it changes temperature. This creates the moderating effects that large bodies of water have on climate and helps ectothermic animals maintain a body temperature that is life sustaining.

Water is also a **lubricant**. Its abundance, combined with its properties of cohesion, solvation, and heat capacity, makes water an extremely valuable lubricant in the bodies of organisms. An obvious example of this is the water in saliva moistening the food and lubricating the esophagus during swallowing.

Water has some other interesting and valuable properties, such as the fact that it is **transparent**. Algae rely on the penetration of light to a considerable depth so they can conduct photosynthesis. Water also has its greatest **density** at 4°C. The way the H-bonds change as water molecules freeze allows the molecules to be closest together at a temperature slightly above freezing. This is why ice floats. A layer of ice on a lake creates an insulation barrier for the aquatic organisms. As temperatures change, the water in the lake will circulate vertically causing spring and fall overturns, which increases the availability of nutrients to organisms.

A.1 CONCEPT CHECK-UP QUESTIONS:
1. What is an organic molecule?
2. Describe a polar covalent bond.
3. What causes H-bonding? What is the effect of H-bonding in water?
4. Why are water's solvation abilities important to life?

ACIDS, BASES, pH, AND BUFFERS

Acids are molecules that dissociate to release hydrogen ions. Hydrochloric acid, HCl, is a strong acid because it readily dissociates producing free H^{1+} and Cl^{1-}. Stomach juices are very acidic because the stomach produces HCl, which it releases into the stomach to aid the digestion of protein.

$$HCl \longrightarrow H^{1+} + Cl^{1-}$$

Bases are molecules that release OH^{1-} ions. Molecules like NaOH are bases.

$$NaOH \longrightarrow Na^{1+} + OH^{1-}$$

Bases have a neutralizing affect on acids. When equal concentrations of acids and bases are combined, their ions will neutralize each other. The H^{1+} from the acid combines with the OH^{1-} from the base and forms water. The other ions form a salt. An example of this is the mixing of HCl and NaOH.

$$H^{1+} + Cl^{1-} + Na^{1+} + OH^{1-} \longrightarrow H_2O + NaCl$$

Water is neither an acid nor a base because it releases both hydrogen and hydroxide ions in equal numbers ($H_2O = H^{1+} + OH^{1-}$), which cancel out the effect of each other. Water is neutral.

pH is a measure of the amount of free H^{1+} in a system. Mathematically, pH is the $-\log[H^{1+}]$. pH numbers can range from a low of 0 to a high of 14. When $[H^{1+}] = [OH^{1-}]$, the pH = 7, which is neutral. Below 7 is acidic (more H^{1+}) and above 7 is basic (more OH^{1-}).

Maintaining pH is critical for organisms. In humans, the pH of blood normally varies from about 7.35 in muscle tissues to 7.38 in the lungs, meaning that blood is slightly basic. In comparison, digestive juices in the stomach have a pH of about 2.5. Large deviations in the pH of the various fluids in the body can affect the shape and function of required enzymes. Without proper enzymes, the tissues and organs can cease to function.

To prevent significant changes in pH and to maintain homeostasis, organisms produce **buffers**. Buffers are molecules that can either pick up or release hydrogen (or hydroxide) ions in order to help maintain the appropriate pH. A very common buffer in the human body is the bicarbonate ion (HCO_3^{1-}), which has significant functions in different systems like the digestion and respiration. Bicarbonate ions are **amphoteric**, meaning they behave like an acid (releasing hydrogen ions) or a base (gaining H^{1+}) to maintain a steady pH.

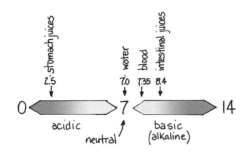

Figure A.2. pH Scale. *pH is a number between 0 and 14. It is calculated from the concentration of H^{1+} in a system (solution). A range of pH values exist in the human body.*

A.2 CONCEPT CHECK-UP QUESTIONS:
1. What is the fundamental difference between an acid and a base?
2. What makes water "neutral"?
3. Is a solution with a low pH more, or less acidic that one with a high pH?
4. What is the role of bicarbonate ions in the blood?

MACROMOLECULES

Water plays a large role in the chemical reactions of cells. When smaller molecules are combined, water is often a product. This type of reaction is called **dehydration synthesis** or **condensation**. The reverse, involving the addition of water, is called **hydrolysis**. Hydrolysis and dehydration synthesis are enzymatic reactions. In some cases, the same enzyme will control the rate of both reactions depending on the conditions.

All four of the major types of macromolecules in living systems form large molecules made out of **monomers** (or **unit molecules**). If the macromolecule is a sequence of many monomers, it is known as a **polymer**. The process of dehydration synthesis builds polymers, where hydrolysis breaks polymers down into their component monomers. The macromolecules of carbohydrates, proteins, and nucleic acids are polymers. Lipids do not really form polymers because the number of unit molecules that can join together to form a lipid macromolecule is limited.

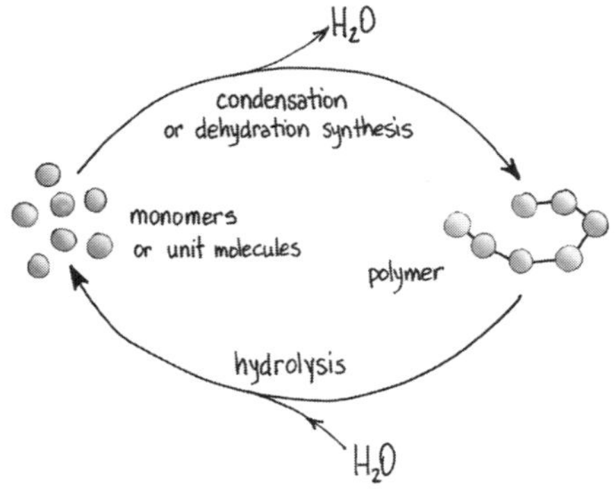

CARBOHYDRATES

Carbohydrates are, literally, hydrates of carbon. The **empirical formula** of a carbohydrate is CH_2O (i.e., $C + H_2O = CH_2O$). Generally speaking, carbohydrates are either sugars or combinations of sugars. **Glucose** ($C_6H_{12}O_6$), the product of photosynthesis is a very common **monosaccharide** (**single** or **simple sugar**). It undergoes dehydration synthesis requiring the enzyme **maltase** to become the **disaccharide** (double sugar) called **maltose**. Continued dehydration synthesis will result in larger molecules called **oligosaccharides** (with a few sugars) and finally **polysaccharides** (containing many sugars).

Examples of other monosaccharides are fructose and galactose. Fructose, galactose and glucose are all **isomers** of each other. This means that they have the same chemical formula ($C_6H_{12}O_6$), but differ from each other in the arrangement of their atoms. Another well-known monosaccharide is deoxyribose ($C_5H_{10}O_4$), which is present in DNA. Note that sugar names end with "ose".

Common disaccharides include **sucrose** or table sugar (glucose + fructose) and lactose, the sugar in milk (glucose + galactose). Polysaccharides differ from one another by the way the monosaccharides are bonded together. These differences are easily noted by examining the three common polymers of glucose. **Starch** is a linear polymer that spirals to form a helix that sometimes branches. **Glycogen**, in comparison, is highly branched, yet **cellulose** is a linear sequence of glucose molecules.

Plants make glucose (via **photosynthesis**), metabolize it for energy, store it as starch and convert it into cellulose to build new cell walls for growth. Our food contains carbohydrates. Starches and disaccharides are digested into monosaccharides (primarily glucose molecules), which are transported by our circulatory system to body cells where they are oxidized in the mitochondria during the process of **cellular respiration**. Excess glucose is stored in the liver in the form of glycogen. People are unable to digest the cellulose components of food. It is sometimes referred to as **dietary fiber** and serves a useful purpose in the production of feces. Glucose and polymers of glucose are obviously very significant to organisms.

Figure A-3. Synthesis and Hydrolysis. *Monomers are combined by dehydration synthesis reactions to form polymers. Water is a byproduct of these reactions. Polymers can undergo hydrolysis, with the addition of water, to reform the monomers.*

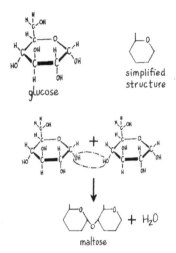

Figure A-4. Glucose and the Formation of Maltose. *Glucose commonly exists as a ring-shaped molecule as illustrated. Dehydration synthesis of two glucose molecules results in the formation of maltose. Water is a byproduct of this reaction.*

A.3 CONCEPT CHECK-UP QUESTIONS:
1. What is the difference between dehydration synthesis and hydrolysis?
2. Why do lipids not form true polymers?
3. Water is a by-product of condensation reactions. Where does it come from?
4. How does the use of glucose in a plant differ from its function in an animal?

LIPIDS

Biochemicals that do not mix freely with polar solvents like water are called **lipids**. There are lots of types and examples of lipids. Next to carbohydrates, lipids are the second most important energy molecules for us. As a category of biochemicals, lipids include fatty acids, neutral fats, oils, steroids like cholesterol and some hormones, waxes and other specialized molecules.

Fatty acids are **non-polar** chains of carbon and hydrogen with a carboxylic acid end. A tremendous number of variations of fatty acids exist. For example, some fatty acids are **saturated** (filled with hydrogen; without double bonds) while others are **unsaturated** (with double bonds, therefore having less hydrogen). In general, the fatty acids produced by animal tissues are more saturated and tend to be solid at room temperatures. Those produced by plant tissues are unsaturated and tend to be liquid, like vegetable oil. A "**trans-fat**" is an unsaturated fatty acid with the **trans conformation**. In terms of nutrition, saturated fatty acids and trans-fats are the least desirable fatty acids to ingest.

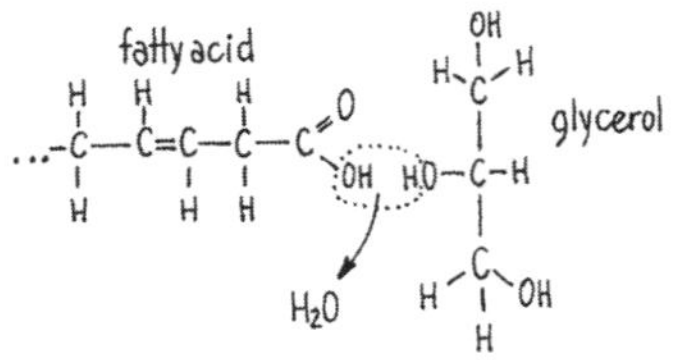

Figure A-5. Formation of a Neutral Fat. *When a fatty acid bonds by dehydration synthesis to a molecule of glycerol, a monoglyceride results. If two or three fatty acids were to bond onto glycerol, diglycerides or triglycerides are formed (respectively). Mono-, di- and triglycerides are neutral fats.*

Neutral fats are produced by the dehydration synthesis of one or more fatty acids with an alcohol, commonly **glycerol**, resulting in the formation of **glycerides**. Many types of glycerides are possible both because of the number and variety of fatty acids that can be incorporated into the structure and because of the three different bonding locations on glycerol for fatty acids. A **monoglyceride** has one fatty acid combined with glycerol, a **diglyceride** has two, and a **triglyceride** has three. Neutral fats are very useful in our bodies where they provide insulation from temperature changes, and damage. They are also an alternate supply of molecules that can be converted into ATP. Unfortunately, however, we tend to store them in **adipose** (fat) cells.

Specialized variations of lipids that are important biologically include phospholipids, steroids and waxes. **Phospholipids** are a variation of a triglyceride where one of the fatty acids is replaced with a phosphate and nitrogen-containing group. This creates a polar region on the otherwise non-polar molecule. Consequently phospholipids can mix with both polar and non-polar materials. Phospholipids are very important to cells, as they are an integral component of all membranes. **Steroids** are non-polar ring structures. Sex hormones (estrogen, testosterone etc.) are steroids. So is cholesterol, which is synthesized by liver cells and transported in the bloodstream. Cholesterol helps maintain the integrity of cell membranes throughout the body. Cholesterol levels in blood must be kept in check to remain healthy. **Waxes** (like earwax) are combinations of fatty acids with alcohols larger than glycerol.

A.4 CONCEPT CHECK-UP QUESTIONS:
1. What is the defining characteristic of lipids?
2. What differences exist between fatty acids from animal tissues and fatty acids from plant tissues?
3. List the ways glycerides can be different from each other.
4. Name two types of lipids that do not contain glycerol.

PROTEINS

Proteins are polymers of unit molecules called **amino acids**. Some proteins are relatively short, like oxytocin with nine amino acids. Insulin is fifty-one amino acids long. Most proteins, however, have over one hundred amino acids in their structure.

The roles of proteins in cells can be described as either "functional" or "structural". Functional proteins include enzymes, antibodies, some hormones, and transport proteins (like hemoglobin). They each have a specific metabolic function. For example, enzymes catalyze chemical reactions. **Maltase** catalyzes the formation of maltose from glucose molecules as well as the hydrolysis of maltose into glucose molecules. **Carbonic anhydrase**, an enzyme in blood, plays a role in preparing carbon dioxide for transport from tissues to the lungs.

Structural proteins, on the other hand, form parts of structures. **Keratin** and **collagen** are examples of these types of proteins. Keratin is a component of fingernails and hair.

Collagen is in skin and connective tissue where it provides extracellular support.

There are twenty different amino acids. All have a common structure. At one end (or terminus) there is an **amino group** and at the other end there is a **carboxyl group**. The carboxyl group denotes a carboxylic acid - hence the name "amino acid". Each amino acid has an "R" group bonded to the central carbon. The difference between the R-groups (acidic, polar, cyclic, etc.) provides each amino acid with a unique chemical nature. This is a significant feature for combinations of amino acids because the resulting protein will have distinct chemical regions.

Two amino acids bond together by dehydration synthesis to form a **dipeptide**. The new bond linking them together is a **peptide bond**. Peptide bonds are strong covalent bonds. A dipeptide has one peptide bond and it holds two amino acids together. A **tripeptide** has two peptide bonds and holds three amino acids together and so on. In this manner, long sequences of amino acids are built.

There are four levels of consideration to the structure of proteins. The first level, or **primary structure**, is simply the sequence of amino acids. Because there are twenty amino acids, it is easy to realize that there are literally millions of different sequences of amino acids. Consequently, millions of different proteins are possible.

As a cell builds a protein one amino acid at a time, the molecule begins to twist due to the angle of the peptide bonds. Because of the attraction between H-bonds, the linear arrangement can take on other **conformations** (shapes). It commonly forms a spring-like shape called an **alpha helix** ($\propto$-helix), which is held together by H-bonds between every fourth amino acid [as oxygen (δ^-) from one amino acid gets close to hydrogen (δ^+) from the other amino acid]. An alternate structure forms when the H-bonds form between linear segments of amino acids forming what is called a **beta (ß)-pleated sheet**. These aspects or levels of protein structure are called **secondary structure**. All proteins have secondary structure.

As amino acids are combined to make proteins, some amino acids, by virtue of their unique chemistry (R-group), cause kinks or bends in the otherwise regular spiral pattern. When this happens, segments of the helix are brought close to each other and new bonds (hydrogen, ionic, and/or covalent) form between these adjacent sections holding it into a three-dimensional shape often described as a bent and folded $\propto$-helix. This third level of structure is known as **tertiary** structure. Many proteins (like enzymes) have tertiary structure.

The hydrogen bonds that contribute to maintaining tertiary structure are fairly easily broken. They are very sensitive to things like pH changes, temperature changes, and the presence of heavy metal ions. Proteins, when exposed to these variations in environmental conditions, may lose their tertiary shape due to the breaking of the hydrogen bonds allowing the protein's shape to alter. Reconfigured proteins like these ones are called **denatured**. This is what happens when milk sours. The protein in milk, casein, denatures and forms insoluble floating lumps. Shape changes cause function changes. In the case of enzymes, an altered tertiary structure means the enzyme will no longer be able to function, as it otherwise would have.

The fourth level of protein structure is called **quaternary structure**. It occurs in some proteins where different tertiary configurations associate together and function as a unit. Hemoglobin is a well-known protein that is actually made up of four tertiary polypeptides associated with a central iron-containing component, called a "heme" group.

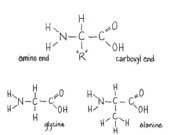

Figure A-6. Amino Acid Structure. *All amino acids have a backbone of N - C - C with an amino terminus at one end and a carboxyl terminus at the other end. There are 20 variations of the "R-group". Two different amino acids are illustrated here.*

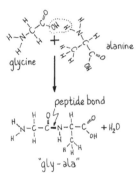

Figure A-7. Formation of a Dipeptide. *Glycine and alanine are illustrated bonding together by a peptide bond to form the dipeptide "gly-ala". Notice that water is a byproduct of this dehydration synthesis reaction. The dipeptide is actually a bent molecule. This bending continues with each additional amino acid and results in the helical shape of secondary structure.*

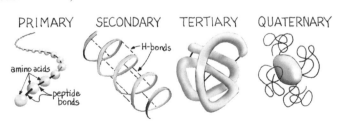

Figure A-8. Levels of Protein Structure. *The primary level is represented as a sequence of amino acids. Secondary structure is the alpha helix resulting from bond angles when the amino acids bond together. All proteins have primary and secondary structure. Tertiary structure is a 3-dimensional shape. Many proteins (like enzymes) have tertiary structure. Quaternary structure is described as the association of tertiary structures. Some proteins like hemoglobin have this level of complexity to the structure.*

A.5 CONCEPT CHECK-UP QUESTIONS:
1. Describe the common structure of amino acids.
2. What type of bond is a peptide bond? Where can one be found?
3. What are the two types of secondary structure in proteins? What causes the formation of each?
4. What type of bonds and what level of structure is destroyed when a protein is denatured?

NUCLEIC ACIDS

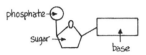

Figure A-9. Nucleotide Structure. *All nucleotides have a common structure. They have a phosphate group, a sugar, and a nitrogenous base. In a DNA nucleotide, the sugar is deoxyribose and the base could be adenine, guanine, cytosine or thymine. For an RNA nucleotide, the sugar is ribose and the bases could be the same, except that uracil takes the place of thymine.*

The final group of biochemicals is **nucleic acids**. There are two different types of nucleic acids, **deoxyribonucleic acid** (**DNA**) and **ribonucleic acid** (**RNA**). Nucleic acids are polymers of unit molecules called **nucleotides**. Nucleotides, themselves are composed of smaller molecules: a sugar, a phosphate group, and a nitrogen-containing (or nitrogenous) **base**. In the case of DNA, the sugar is **deoxyribose**. In RNA, it is **ribose**. The phosphate group is always the same, but the bases differ slightly between DNA and RNA.

DNA is a **double helix**. It has a spiraling ladder-shaped backbone of alternating deoxyribose and phosphate molecules. Bases are bonded onto each sugar and H-bonded to each other holding the two strands together. There are four different bases in DNA. The base **adenine** is a double ring structure called a **purine**. It is always paired up with **thymine** (a smaller, single-ringed structure called a **pyrimidine**). Similarly, **guanine** is always H-bonded to **cytosine** (purine with pyrimidine, again). Other combinations of these purines and pyrimidines do not exist because they are unable to H-bond to each other. In contrast, RNA is a shorter, single stranded molecule produced from a segment of DNA. The base thymine does not exist in RNA. It is replaced by **uracil**, which is a very similar pyrimidine.

Table A-1. Contrasts between DNA and RNA

Basis of Contrast	DNA	RNA
Sugar	Deoxyribose	Ribose
Location	Nucleus	Nucleus and cytoplasm
# strands	Two	One
Bases Present	A, G, T, C	A, G, U, C

DNA, along with proteins called histones, forms chromosomes. This genetic molecule has three functions. It must be able to replicate, or make copies of itself for cell division. It also controls cellular activity by containing the codes (blueprints) for the synthesis of all proteins, including enzymes. DNA must also be able to change, or mutate, to provide raw material for Natural Selection. The functions of DNA are the topic of Unit C.

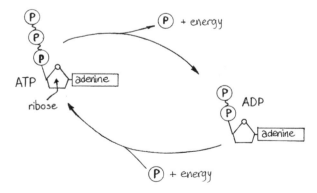

Figure A-10. ATP and Cellular Energy. *ATP is a specialized ribose nucleotide that has two extra phosphate groups attached to it by high-energy covalent bonds. These bonds can break and reform relatively easily as cells meet their energy requirements. When ATP undergoes hydrolysis, it loses a phosphate and becomes ADP. The energy from the P~P bond is available for cell processes. When energy is available, ATP can be reformed and the energy is stored again.*

ATP

One particularly important nucleic acid is the modified nucleotide known as adenosine triphosphate (**ATP**). ATP is an RNA nucleotide with an adenine base (adenine + ribose = adenosine) with two additional phosphate groups attached to it.

The phosphate~phosphate bonds are unique in that they are weak, but very energy-rich bonds. Cells store energy in this way - as chemical bond energy. In order to release the energy, an enzyme, **ATPase**, breaks the outer P~P bond in ATP, thus producing **ADP** and releasing a phosphate group. Once sufficient energy is available in the cell, the components of ATP can be reunited back into the energy storage molecule. The addition of a phosphate group to a molecule, such as what happens during the synthesis of ATP is called **phosphorylation**. This combination of reactions is often illustrated as the **ATP Cycle**. It is easy to see why ATP is often called the energy currency of a cell (because cells "produce" and "spend" ATP as they save and use the energy of the P~P bonds).

A.6 CONCEPT CHECK-UP QUESTIONS:
1. How do the components of a DNA nucleotide differ from those of an RNA nucleotide?
2. What's the difference between a purine and a pyrimidine?
3. What two types of biochemicals make up chromosomes?
4. Describe the bonds between the phosphates in ATP. What is their function?

CHECK YOUR UNDERSTANDING

MULTIPLE CHOICE:

1. Water is all of the following except
   A. Bent.
   B. Polar.
   C. Organic.
   D. Covalent.

2. The polarity water molecules is due to
   A. molecular cohesiveness.
   B. their ability to act as a solvent.
   C. H-bonding to other molecules.
   D. the unequal sharing of electrons.

3. What type of bond holds water molecules to each other?
   A. Ionic.
   B. Covalent.
   C. Adhesion.
   D. Hydrogen.

4. Hydrogen bonds form between
   A. O atoms of two molecules.
   B. H atoms of different molecules.
   C. H and O atoms within a molecule.
   D. H of one molecule and the O of another.

5. The most abundant molecule in the human body is
   A. organic and ionic.
   B. inorganic and ionic.
   C. organic and covalent.
   D. inorganic and covalent.

6. The gain of electrons due to unequal electron sharing causes a
   A. low pH.
   B. high pH.
   C. positive dipole.
   D. negative dipole.

7. The water in an organism's body helps
   A. reactions occur.
   B. transport molecules.
   C. maintain body temperature.
   D. All of the above choices are true.

8. Which is **NOT** a function of water in the human body?
   A. Hair growth.
   B. Food digestion.
   C. Urine excretion.
   D. Nutrient distribution.

9. Pure water is neutral because it contains
   A. only H^{1+}.
   B. only OH^{1-}.
   C. equal amounts of H^{1+} and OH^{1-}.
   D. seven times more H^{1+} than OH^{1-}.

10. Which is closest to the pH of blood?
    A. 4.8
    B. 6.7
    C. 7.3
    D. 10.8

11. The higher the pH value, the
    A. lower the pH number.
    B. more basic the solution.
    C. more acidic the solution.
    D. greater the H^{1+} concentration.

12. Which **BEST** describes an acid?
    A. Lots of H^{1+} present.
    B. Lots of OH^{1-} present.
    C. More H^{1+} than OH^{1-} present.
    D. More OH^{1-} than H^{1+} present.

13. Buffers that keep the pH at 8.3 **MOST LIKELY**
    A. release H^{1+}
    B. bond with H^{1+}
    C. bond with OH^{1-}
    D. dissociate to release H^{1+} and OH^{1-}

14. KOH dissociates into K^{+1} and OH^{-1} and is therefore
    A. a base.
    B. an acid.
    C. a neutral substance.
    D. a solvent for non-polar substances.

15. Which statement is correct regarding acids and bases?
    A. They combine to form buffers.
    B. Acids lower pH; bases raise pH.
    C. Acids are harmful but bases are not.
    D. Acids combine with H^{+1}; bases combine with OH^{-1}.

16. If a base were added to a solution buffered at a pH of 6.0, the pH would
    A. drop.
    B. not change.
    C. rise sharply.
    D. increase slightly.

17. Which pair is mismatched?
    A. Blood; pH = 7.35
    B. Gastric juice; pH = 2.5
    C. Salivary juice; pH = 10.3
    D. Pancreatic juice; pH = 8.4

18. Which equation shows synthesis?
    A. acid + base ▶ salt + water
    B. ATP ▶ ADP + P + energy
    C. 2(glucose) ▶ maltose + water
    D. protein + water ▶ amino acids

19. Hydrolysis occurs when
    A. glucose forms maltose and water.
    B. amino acids + water form dipeptides.
    C. RNA breaks into nucleotides and releases water.
    D. fats + water form glycerol + fatty acids.

20. Which type of biochemical has the empirical formula: CH_2O ?
    A. Lipid.
    B. Protein.
    C. Nucleic acid.
    D. Carbohydrate.

21. Which bond type exits between the bases in DNA?
    A. Ionic.
    B. Peptide.
    C. Covalent.
    D. Hydrogen.

22. Which is **NOT** a polysaccharide?
    A. Starch.
    B. Glycerol.
    C. Cellulose.
    D. Glycogen.

23. Converting saturated fatty acids into unsaturated fatty acids requires
    A. adding H and double bonds.
    B. removing H and double bonds.
    C. removing H and adding double bonds.
    D. adding H and removing double bonds.

The next question refers to this diagram.

24. What product forms by the repetition of this process?
    A. Tripeptide.
    B. Triglyceride.
    C. Trinucleotide.
    D. Trisaccharide.

25. Which has an empirical formula of CH_2O?
    A. ATP.
    B. DNA.
    C. Ribose.
    D. Glycerol.

26. The hydrolysis of a neutral fat results in
    A. glycerol and fatty acids only.
    B. glycerol, fatty acids and water.
    C. amino acids and fatty acids only.
    D. amino acids, fatty acids and water.

27. Which biochemical forms membranes?
    A. Hormones.
    B. Triglycerides.
    C. Phospholipids.
    D. Disaccharides.

28. What are phospholipids made of?
    A. cholesterol, glycerol, fatty acids.
    B. phosphate, cholesterol, protein.
    C. fatty acids, phosphate, glycerol.
    D. glycerol, amino acids, phosphate.

29. Which is **NOT** a role of neutral fats?
    A. Cushion tissues from damage.
    B. Offset the effect of pH changes.
    C. Insulate against temperature loss.
    D. Store energy for ATP production.

30. The rigidity and strength of plant cell walls is due to
    A. cellulose.
    B. glycogen.
    C. cholesterol.
    D. phospholipid.

31. Which of these molecules is the most highly branched?
    A. Starch.
    B. Maltose.
    C. Glucose.
    D. Glycogen.

32. How many peptide bonds link 108 amino acids together?
    A. 106
    B. 107
    C. 108
    D. 109

33. When two amino acids join together, they form a
    A. dipeptide.
    B. diglyceride.
    C. dinucleotide.
    D. disaccharide.

34. An alpha helix bent and folded into a 3-D shape represents
    A. primary structure.
    B. secondary structure.
    C. tertiary structure.
    D. quaternary structure.

The next question refers to this diagram.

35. What type of bond is bond **X**?
    A. Ionic.
    B. Peptide.
    C. Hydrogen.
    D. Electrostatic.

36. Denaturing protein results in the loss of
    A. primary structure.
    B. secondary structure.
    C. tertiary structure.
    D. quaternary structure.

37. The helical portion of a polypeptide is
    A. a sequence of "R" groups only.
    B. ..RCC..RCC..RCC..RCC.. etc.
    C. ..NCC..NCC..NCC..NCC.. etc.
    D. ..P – sugar..P – sugar..P – sugar..etc.

38. RNA is composed of a series of
    A. purines.
    B. nucleotides.
    C. pyrimidines.
    D. amino acids.

39. Which combination is a nucleotide?
    A. base – acid – salt
    B. DNA – RNA – water
    C. base – sugar – phosphate
    D. adenine – thymine – uracil

40. What is **FALSE** about RNA and DNA?
    A. RNA has uracil; DNA does not.
    B. RNA sugars have more oxygen than DNA sugars.
    C. RNA is single stranded where DNA is double stranded.
    D. RNA can only be found in the cytoplasm where DNA can only be found in the nucleus.

The next question refers to this diagram.

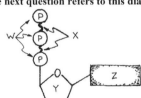

41. "Available energy" for cells is represented by
    A. W
    B. X
    C. Y
    D. Z

42. ATP is significant in cells because it
    A. contains adenine
    B. stores ribose for RNA.
    C. has high-energy bonds.
    D. is the unit molecule of DNA.

43. ATP is **BEST** described as a specialized
    A. RNA nucleotide with a purine base.
    B. DNA nucleotide with a purine base.
    C. RNA nucleotide with a pyrimidine base.
    D. DNA nucleotide with a pyrimidine base.

WRITTEN ANSWERS:

1. a. Describe the structure of a water molecule.
    b. How does this structure contribute to its unique properties?
    c. Name three ways that water is important *in a cell.*

2. Compare an acid and a base. If the concentration of hydrogen ions in a system were to increase, would pH increase or decrease? Explain.

3. Using glucose and maltose, explain the relationship between synthesis and hydrolysis. Include the role of water in your answer.

4. For each of the four classes of biochemicals:
    a. Identify the unit molecules.
    b. Name at least **ONE** cellular function.

5. Use examples to distinguish between a monosaccharide, a disaccharide, and a polysaccharide.

6. Why is the chemical formula of maltose not a multiple of its empirical formula?

7. Design a data table to contrast three different polysaccharides in three different ways. Complete your table.

8. Draw and label the structure of an amino acid. How do different amino acids differ from one another?

9. Design a data table that will allow you to contrast the four levels of protein structure in terms of their shape and the bonding that holds them together. Complete your table.

10. How does RNA differ from DNA? How are they similar?

11. Draw a molecule of ATP. What is its function and how is it designed for this particular function?

12. Contrast each of the following pairs:
    a. ribose vs. deoxyribose
    b. saturated vs. unsaturated
    c. carboxyl group vs. amine group

13. Explain the significance of the these:
    a. glycine
    b. glycerol
    c. glucose
    d. glycogen
    e. guanine

14. Define and give an example of these:
    a. buffer
    b. dissociate
    c. polypeptide
    d. helix
    e. cohesion.

UNIT B - CYTOLOGY

Cytology, the study of cells, had its birth with the invention of the microscope in the 17th century. In the years that followed, this new technology was improved and scientists became increasingly familiar with the microscopic features of cells. In the late 1830's two German biologists, Schleiden (a botanist) and Schwann (a zoologist) independently observed that all the organisms they were studying were composed of cells. This realization is now generalized as "the cell is the building block of organisms." A few years later, Rudolph Virchow added that "cells come from pre-existing cells" based on his observations. These ideas comprise what is now known as the **Cell Theory**.

As the scientific study of cells progressed, scientists developed a classification system for cells based on their complexity. **Prokaryotic** cells, like bacteria, are quite simple because they do not have any membranous **organelles** such as a nucleus, mitochondria, and chloroplasts. Animal, plant and fungus cells are **eukaryotic** cells and contain organelles with membranes. Each organelle has its own characteristic structure, function and contribution to the overall function of the cell.

The biggest advantage eukaryotic cells have over prokaryotic cells is **compartmentalization**. With an assortment of membrane-bound spaces (organelles), countless different reactions and processes can co-exist in eukaryotic cells without interfering with one another. Thus eukaryotic cells are more efficient, diversified and specialized. This unit considers the structure and function of many of the organelles that contribute to compartmentalization.

NUCLEUS AND NUCLEOLUS

The **nucleus** has been called a cell's "control centre". It is a eukaryotic cell's most obvious, visible organelle. A membrane called the **nuclear envelope** (or **nuclear membrane**) encloses it. The nuclear envelope has **pores** in it that make the fluid of the nucleus, called the **nucleoplasm** ("nucleus fluid"), somewhat continuous with the **cytoplasm** ("cell fluid"). These pores allow large molecules to pass between the nucleus and the cytoplasm.

The genetic material known as **chromatin** is suspended in the nucleoplasm. Chromatin is made out of DNA and protein. It becomes known as **chromosomes** when it is coiled up for cell division. In its unraveled form, segments of the chromatin produce RNA for manufacturing of proteins in the cytoplasm. RNA is an example of a nuclear product that leaves the nucleus through nuclear pores. The production of RNA (a process called **transcription**) requires enzymes, which are specialized proteins produced at ribosomes in the cytoplasm. Enzymes enter the nucleus through nuclear pores. It is through this type of molecular communication with the cytoplasm that a nucleus controls cellular activities.

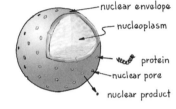

Figure B-1. The Roles of Nuclear Pores. *A double membrane sometimes referred to as the nuclear envelope surrounds a nucleus. This membrane has pores in it for the inward passage of large molecules (like proteins) and the outward passage of large nuclear products (like RNA).*

The **nucleolus** is a visible region within the nucleoplasm. It is here that a type of RNA (called ribosomal RNA) and specific proteins are combined to form the sub-units of the non-membranous structures called **ribosomes**. A complete ribosome is made out of two unequal-sized sub-units, which leave the nucleus through the pores and get assembled into ribosomes in the cytoplasm. Some ribosomes in eukaryotic cells become embedded in the membranes of the **endoplasmic reticulum** (ER) where others remain free in the cytoplasm. Ribosomes are the site of protein synthesis. It is on their surfaces that the amino acids are individually put into their predetermined sequences to form protein molecules. **Polysomes** are clusters of chemically identical ribosomes that allow the almost simultaneous mass production of particular proteins. Proteins produced by free ribosomes in the cytoplasm are generally used within the cell (e.g. enzymes that catalyze reactions). Those produced by ribosomes embedded in the ER are often secreted from the cell.

B.1 CONCEPT CHECK-UP QUESTIONS:
1. What is the advantage of compartmentalization in cells?
2. Give two specific examples that illustrate the function of nuclear pores.
3. Describe a ribosome. Where are the subunits of ribosomes produced?
4. What is the relationship between a ribosome and a polysome?

ENDOPLASMIC RETICULUM

The endoplasmic reticulum is a set of membranous channels that extend through a cell's cytoplasm. There are two types of ER: rough ER and smooth ER. The **rough endoplasmic reticulum** (RER) has ribosomes embedded in its membranes. Ribosomes are the site of protein synthesis. The proteins manufactured at the ribosomes of the RER enter the **intramembranous space** within the RER. The **smooth endoplasmic reticulum** (SER) contains enzymes for the synthesis of needed biochemicals. For example, cells that produce steroids, such as the hormone-producing cells of the testes, have abundant SER. The SER also plays a role in the synthesis of membrane phospholipids and removing any toxins that accumulate in the cytoplasm.

Many illustrations and electron micrographs show the SER and RER as discrete organelles. Others show them as interconnected, which is most likely, as the functions of SER require enzymes. An interconnection with RER would provide the communication link and route for the movement of the enzymes (proteins produced by the ribosomes of RER) to the SER. Other illustrations suggest that the ER is also connected to the nuclear envelope, which would help move RNA and ribosomal sub-units out of the nucleus and onto RER where they are both used in protein synthesis.

Both types of ER serve a transportation function in the cell by facilitating the movement of molecules from one area to another. Both accumulate molecular products in their interiors. Where SER accumulates the steroids they produce, RER accumulates the proteins produced by the ribosomes that are embedded in them. Sections of ER containing these products then break free in a process often described as **"blebbing"** or "budding" to produce small membrane bound sacs called **vesicles** enclosing the biochemicals. These products are often destined for secretion from the cell. The vesicles are called **transition** (or transport) **vesicles**. SER may be more involved with the formation of transition vesicles, thus proteins from the RER would have to travel to it to be packaged for eventual secretion.

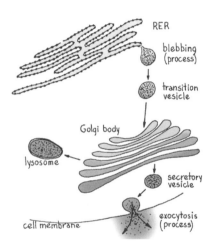

Figure B-2. The Secretory Pathway. *This sequence of events is an excellent example of the inter-relationships between cell organelles. ER produces a vesicle containing its products (proteins for RER, lipids for SER). This vesicle moves to the Golgi body, which modifies the contents and prepares them for secretion by putting them in a second vesicle that joins the cell membrane and exocytosis occurs. In another inter-relationship the proteins in the vesicle from the RER are activated as hydrolytic enzymes by the Golgi bodies and used for the formation of a lysosome.*

GOLGI BODIES

ER cannot function efficiently in cells without the presence of another membranous structure called **Golgi apparatus** or **Golgi bodies**, (named after Camillo Golgi the Italian biologist that discovered them in the late 1800's). Golgi bodies may look similar to smooth ER at first glance. They are, however, comprised of a set of several discrete, flattened but curved **saccules** usually located between ER and the cell membrane.

Golgi bodies receive transition vesicles from the ER. They concentrate the molecular contents of the vesicles, and chemically adjust some of the molecules in preparation for their eventual secretion. New vesicles form by blebbing from the Golgi apparatus. These ones are called **secretory vesicles** and they contain the biochemicals that have been modified by the Golgi apparatus. Secretory vesicles move to and join with the cell membrane in such a way that their contents are expelled from the cell, a process named **exocytosis**. This sequence of functional interrelationships among organelles is known as the **secretory pathway**. A specific example of a reaction that occurs in the Golgi bodies is the addition of carbohydrate chains to proteins thus making molecules called **glycoproteins**, which become components of cell membranes.

The Golgi bodies also produce vesicle-like structures called **lysosomes**, which contain **hydrolytic enzymes** used for the **hydrolysis** (digestion) of molecules in cells. Lysosomes typically have a double outer membrane to give the cell greater protection against the enzymes becoming free in the cytoplasm. These enzymes could potentially destroy a cell.

Unicellular organisms use lysosomes to digest food particles they ingest by **endocytosis**. In this process the food particles become enveloped by a section of the cell membrane and form a structure known as a food **vacuole**. The food vacuole fuses with the lysosome and digestion follows. This is **intracellular digestion** (digestion within a cell) and it results in the production of simple molecules (amino acids, monosaccharides, etc.), which then leave the lysosome and enter the cytoplasm where they are used by the cell. Multicellular organisms use lysosomes for the **autodigestion** (self-digestion) of unwanted cells and cell parts. A well-known example of this the digestion of a tadpole's tail as it undergoes metamorphosis to become a frog.

B.2 CONCEPT CHECK-UP QUESTIONS:
1. What are the structural and functional distinctions between RER and SER?
2. What is the advantage of an RER connection to the: a) nuclear membrane? b) SER?
3. How might the contents of transition vesicles differ from secretory vesicles?
4. What is the source and function of the enzymes in a lysosome?

MITOCHONDRIA AND CHLOROPLASTS

Mitochondria and **chloroplasts** are related both structurally and functionally. Both are fluid-filled and have complex internal membranes. Energy-related chemical reactions occur in their interiors as well as in their internal membranous structures.

Mitochondria exist in almost all cells. Their inner membranes are very folded, looping back and forth producing tremendously large surface areas. This shape produces the shelf-like appearance of the inner membrane, the "shelves" of which are called **cristae**. The fluid within the inner membrane is called the **matrix**. Each mitochondrion conducts **cellular respiration** by oxidizing parts of carbohydrates from the cytoplasm and capturing the energy from the chemical bonds that get broken. These reactions occur at reaction sites within the cristae. It is advantageous to the cell to have mitochondrial membranes with large surface areas so that more reactions can occur. The energy released from these reactions is used to **phosphorylate** (add phosphorous to) ADP, producing ATP. In this way cells are able to retain this energy in the high-energy P~P bonds of **ATP**. With the energy stored in this form, cells can access it easily when needed. The energy released from the complete oxidation of one glucose molecule can be used to generate 38 molecules of ATP. For this reason, mitochondria have been nicknamed "the powerhouses of cells".

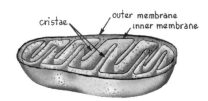

Figure B-3. The Structure of a Mitochondrion. *Every mitochondrion has a double membrane, which separates the inner fluids into two parts: the intra-membranous fluid and the matrix (in the middle). Their inner membrane has tremendous surface area achieved by folding back and forth forming extensions into the matrix called cristae. Cristae are the locations of the reactions that make ATP.*

$$C_6H_{12}O_6 + 6O_2 \longrightarrow 6CO_2 + 6H_2O + 38ATP$$

Chloroplasts are organelles that exist only in some plant cells. They contain pigment molecules like **chlorophyll**, which enable them to capture sunlight energy and use it to build sugar molecules. This process is **photosynthesis**.

$$6CO_2 + 6H_2O \longrightarrow C_6H_{12}O_6 + 6O_2$$

The functional interrelationship between mitochondria and chloroplasts becomes obvious when the net reactions for cellular respiration and photosynthesis are compared. The reactants of one process are the products of the other. Photosynthesis makes glucose, which cellular respiration oxidizes for energy. Some plant cells have mitochondria and some of the sugars plants make are used for energy within their own tissues. Other sugar molecules are converted to starch for storage or into cellulose to make the cell walls needed by the plant as it grows. Herbivores digest and metabolize the plant tissues that they consume. In this way, a plant's carbohydrates (sugars, starch etc.) provide energy for animals.

OTHER CELL FEATURES

CELL WALL

Generally, animal cells are considered to be the only cell type that lacks cell walls. The cell walls of plant cells are made out of the linear polysaccharide, cellulose. As plant cells mature, they deposit cellulose outside of their cell membranes. Where this cellulose cover provides a measure of protection for the inner living parts of the cell, it becomes restrictive because it prevents significant changes in shape and growth of mature cells. It also prevents processes like exocytosis and forces these cells to accumulate their cellular products in large central vacuoles, which add pressure against the cells' organelles. In this way cell walls can contribute to the death of plant cells.

CILIA AND FLAGELLA

Many cells have cilia or flagella that they depend on for some type of movement. Both cilia and flagella have the same structure. They are elongate arrangements of **microtubules** made from the protein **tubulin**. Anchored by structures called **basal bodies**, the microtubules extend through a cell's membrane and into the environment. Basal bodies have a similar structure to cilia and flagella, and are believed to originate from the **centrosome**, an area close to the nucleus (**centrioles** in animal cells). A cilium is much shorter than a flagellum. The outer margins of ciliated cells are covered with cilia that wave in an undulating manner and move fluids over their surface, as in the case of mucus up the trachea. For a *Paramecium*, this action provides its swimming movement. In contrast, flagellated cells are able to swim due to the activity of relatively few numbers of long flagella. A sperm cell, for example, has a single flagellum.

CYTOSKELETON

The organelles in cells do not "float around" freely in the cytoplasm. Cells have a **cytoskeleton**. The cytoskeleton is an array of **microtubules** and other protein filaments called **microfilaments**. Some parts of the cytoskeleton originate from the centrosome (centrioles). Other parts of the cytoskeleton are attached to the inner surface of the cell membrane from where they anchor organelles and provide a framework for their movement. **Actin** is an example of a protein microfilament. It is capable of "contracting" to provide movement by sliding across the surface of other proteins to produce a shortened structure, as in the case of muscle cell contraction.

B.3 CONCEPT CHECK-UP QUESTIONS:
1. What is the advantage of the large surface area of a mitochondrion's inner membrane?
2. What is the source of the energy for the phosphorylation of ADP into ATP by the mitochondria?
3. How are the functions of mitochondria and chloroplasts related?
4. What is a cytoskeleton made out of?

CHECK YOUR UNDERSTANDING

MULTIPLE CHOICE:

1. Preventing ATP production **MOST** likely affects the
   A. ribosomes.
   B. mitochondria.
   C. Golgi apparatus.
   D. rough endoplasmic reticulum.

2. Cells lacking RER would **NOT** likely be able to produce
   A. vesicles.
   B. any proteins.
   C. steroid hormones.
   D. enzymes for secretion.

3. Which organelles make O_2 and use CO_2?
   A. Lysosomes.
   B. Chloroplasts.
   C. Mitochondria.
   D. Endoplasmic reticulum.

4. Which **BEST** describes a function of SER?
   A. Protein storage.
   B. Communication.
   C. Energy distribution.
   D. Intracellular transport.

5. Which organelles are rich in RNA?
   A. ribosomes.
   B. lysosomes.
   C. mitochondria.
   D. transport vesicles.

6. Which **BEST** describes the traffic through a nuclear pore?
   A. Enzymes in; nucleic acids in.
   B. Enzymes out; nucleic acids in.
   C. Enzymes in; nucleic acids out.
   D. Enzymes out; nucleic acids out.

7. Where in a cell are steroids produced?
   A. Nucleus.
   B. Cytoplasm.
   C. Rough endoplasmic reticulum.
   D. Smooth endoplasmic reticulum.

8. One function of Golgi apparatus is
   A. Phagocytosis.
   B. Protein synthesis.
   C. Intracellular digestion.
   D. Modifying products for secretion.

9. Which is **NOT** a role of SER?
   A. Protein synthesis.
   B. Intracellular transport.
   C. Detoxification of cytoplasm.
   D. Making membrane phospholipids.

10. Lysosomes are produced by the
    A. nucleus.
    B. cell membrane.
    C. Golgi apparatus.
    D. endoplasmic reticulum.

11. The connected nature of RER and SER
    may help move
    A. enzymes from RER to SER.
    B. ribosomes into the cytoplasm.
    C. phospholipids from SER to RER.
    D. vesicles from ER to Golgi apparatus.

Use this sketch for the next five questions.

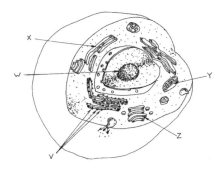

12. What is synthesized by organelle **V**?
    A. ATP.
    B. proteins.
    C. vesicles.
    D. ribosomes.

13. Where are the subunits of ribosomes
    synthesized?
    A. W
    B. X
    C. Y
    D. Z

14. Vesicles that conduct exocytosis are **MOST**
    likely produced by
    A. V
    B. X
    C. Y
    D. Z

15. A function of organelle **Y** is to
    A. make ATP.
    B. synthesize proteins.
    C. manufacture glucose.
    D. prepare secretory vesicles.

16. A function of organelle **Z** is
    A. enzyme storage.
    B. lysosome production.
    C. detoxify the cytoplasm.
    D. steroid manufacture and packaging.

17. When secretory products are being
    transported to the cell membrane, they
    A. move through the ER.
    B. are enclosed in a vesicle.
    C. are transported by carrier molecules.
    D. diffuse freely through the cytoplasm.

18. Which of the following is **FALSE**?
    A. RER helps produce proteins.
    B. Lysosomes can destroy cells.
    C. RER can be modified into SER.
    D. Golgi apparatus packages products.

19. Cells that engulf and digest bacteria could
    be expected to have
    A. nuclei.
    B. nucleoli.
    C. ribosomes.
    D. lysosomes.

20. Which is **NOT** a membranous organelle?
    A. Vesicle.
    B. Ribosome.
    C. Lysosome.
    D. Mitochondrion.

21. The **BEST** sequence of organelles involved
    in the production and secretion of protein is
    A. Ribosome, Golgi apparatus, vesicle, ER,
       cell membrane.
    B. Ribosome, ER, Golgi apparatus, vesicle,
       cell membrane.
    C. ER, ribosome, vesicle, Golgi apparatus,
       cell membrane.
    D. Cell membrane, vesicle, ribosome, ER,
       Golgi apparatus.

22. Genetic information is contained in
    A. Polysomes.
    B. Ribosomes.
    C. Lysosomes.
    D. Chromosomes.

23. What are flagella and cilia composed of?
    A. Lipids.
    B. Proteins.
    C. Nucleic acids.
    D. Carbohydrates.

24. Which sequence of the cell parts listed
    below is most closely associated with
    protein secretion?
    1. vesicle
    2. cell membrane
    3. Golgi apparatus
    4. RER
    A. 4, 1, 2, 1, 3
    B. 2, 1, 3, 1, 4
    C. 4, 1, 3, 1, 2
    D. 4, 3, 1, 2, 1

25. Cells with vesicles could logically have
    A. Golgi apparatus.
    B. Rough endoplasmic reticulum.
    C. Smooth endoplasmic reticulum.
    D. All of these answers could be correct.

26. Which statement is **FALSE**?
    A. Mitochondria make ATP.
    B. Chloroplasts use oxygen.
    C. The nuclear envelope has pores.
    D. Ribosomes contain and make protein.

27. Which organelle stores enzymes?
    A. Lysosome.
    B. Chloroplast.
    C. Mitochondrion.
    D. Golgi apparatus.

28. Which sequence of these cell parts is
    correctly associated with intracellular
    digestion?
    1. food vacuole
    2. cell membrane
    3. lysosome
    A. 2, 1, 3
    B. 1, 2, 3
    C. 2, 3, 1
    D. 1, 2, 3

WRITTEN ANSWERS:

1. Describe how a cell would produce and
   secrete a steroid hormone. Name and
   describe the functions of at least four cell
   structures that are involved.

2. State **ONE** way that the following pairs of
   cell structures <u>are functionally related</u>.
   a. ribosomes and lysosomes.
   b. vesicles and Golgi apparatus.
   c. cell membrane and vesicle.
   d. ribosomes and endoplasmic reticulum.

3. Describe the process of **intracellular
   digestion**. Name the organelles involved
   in this process and state the purpose of it.

4. Contrast each of the following in **ONE** way.
   State the basis of contrast each time.
   a. chloroplast vs. mitochondrion
   b. endocytosis vs. exocytosis
   c. SER vs. RER
   d. vesicle vs. lysosome

5. Study these sketches of organelles.
   a. Name each organelle illustrated.
   b. Describe **ONE** function of each.

UNIT C – DNA AND ITS FUNCTIONS

Genetic information is stored in structures called chromatin, which is made out of DNA combined with small protein molecules known as **histones**. In preparation for cell division, DNA replicates, or makes a copy of itself, thus ensuring the transfer of the correct genetic information to daughter cells. Errors in this process are mutations. After replication, the chromatin coils about the histones becoming more densely packed and visible. These structures are the chromosomes that can be observed microscopically during cell division.

When cells are not dividing, the DNA exists in its unwound form and is actively engaged in controlling cellular activities by governing the synthesis of proteins. To do this, other molecules like RNA and specific enzymes associate with the DNA to communicate its genetic message to the cytoplasm where protein synthesis occurs.

Biotechnology has developed to the point where scientists now manipulate chromatin. During **recombinant DNA** procedures, for example, segments of genetic information are removed from their cells and inserted into host cells producing "hybrid" cells with new characteristics and capabilities.

DNA STRUCTURE

DNA is a double-stranded helical polymer of nucleotides. Each DNA nucleotide consists of three parts: **phosphoric acid**, sugar (**deoxyribose**), and a **base**. There are two types of bases categorized by size. **Purines** are larger, double-ring structures. **Adenine** (A) and **Guanine** (G) are purines. **Pyrimidines** are smaller, single-ring structures. **Cytosine** (C) and **Thymine** (T) are pyrimidines.

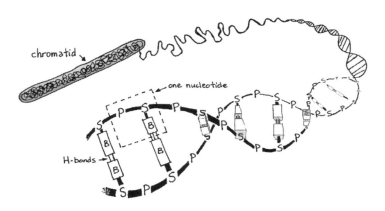

Figure C-1. DNA Structure. *DNA is a double-stranded helical polymer of nucleotides. Hydrogen bonds between the bases hold the strands together. In the condensed form (when the helix wraps up around proteins called histones), DNA strands are known as chromatids.*

Within any DNA molecule, the amount of thymine is always equal to the amount of adenine and, similarly, cytosine and guanine always exist in equal numbers because the molecules are paired with each other and held together by H-bonding. This relationship is called **complementary base paring**, i.e., A is complementary to T and G is complementary to C. Because each base pair is made up of one purine and one pyrimidine, the "sugar - phosphate backbones" of a DNA molecule, even though helical, are parallel. Adenine will not attach to Cytosine, nor will Guanine combine with Thymine even though these combinations are a purine plus a pyrimidine because the molecular dipoles on these pairs of bases do not match each other.

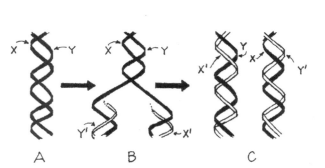

Figure C-2. DNA Replication. *To ensure that daughter cells have a full set of chromosomes, DNA must be able to copy itself accurately. This is the process of replication. To accomplish this, the DNA molecule "unzips" to expose its bases. New nucleotides bond in a complementary fashion to the exposed bases on both sides and then these new nucleotides bond in sequence to form new strands, which wrap around each other. The result is two identical molecules, each with an original strand as well as a new strand.*

REPLICATION

The growth of multicellular organisms and the repair and maintenance of their tissues require cell division. The DNA in body cells must be able to **replicate** (make copies of) itself for cells to divide successfully. Given that daughter cells are supposed to have an exact copy of the parent cell DNA, a parent cell must be able to reproduce its DNA before the daughter cells can be made.

Replication occurs during **interphase**. Although the replication of this huge twisted molecule is very complex, the process can be simplified into these three steps:

1. The enzyme, **DNA helicase** breaks the H- bonds between the complementary bases in DNA allowing the DNA to **unzip** or separate into two strands. This exposes the bonding locations of the bases on each strand.
2. New DNA nucleotides move into the nucleus. **DNA polymerase** enzymes ensure the nucleotides bond onto their exposed complementary bases on the parent strands. This lines up the new set of nucleotides.
3. DNA polymerase then promotes the sequential bonding of these new nucleotides to each other forming new strands of DNA. This process continues along the length of the DNA and results in two separate identical DNA molecules, each with an original strand as well as a new strand. This feature of replication has earned it the phrase **"semi-conservative"**.

C.1 CONCEPT CHECK-UP QUESTIONS:
1. Besides nucleic acids, what type of biochemical is part of a chromosome?
2. What feature helps keep the two stands in a DNA molecule parallel to each other?
3. What would be the outcome if replication did not occur properly prior to cell division?
4. Contrast the roles of DNA helicase and DNA polymerase during replication.

PROTEIN SYNTHESIS

A second function of DNA that cells routinely engage in is overseeing the assembly of amino acids into proteins at the **ribosomes**. Ribosomes can either be embedded in the walls of **endoplasmic reticulum** or located free in the cytoplasm. The difference between these two protein production sites corresponds to the eventual destination of the protein. If the protein is to be used for cell structures or metabolic activities in the cytoplasm, then it is manufactured at a ribosome that is free in the cytoplasm. Those made at RER usually follow the secretory pathway and are exported via **exocytosis**. A third possibility is in the case of hydrolytic enzymes, which are made at the RER and eventually packaged into lysosomes by the Golgi bodies where they are used for intracellular digestion.

No matter which type of protein is being considered, they are all produced in the same way with a two-part process: **transcription** followed by **translation**. Transcription involves the "re-writing" of the genetic information of DNA into **messenger RNA (mRNA)**, which is a chemical version of the gene that can leave the nucleus. During translation, mRNA is used at the ribosomes to assemble amino acids into proteins.

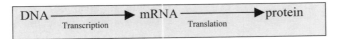

The complex nature of protein synthesis deserves close examination. When a cell is promoted to start making a protein, a **gene** (section of DNA) "puffs up" through the breaking of the H-bonds holding the two strands of DNA together. Similar to replication, this exposes the base sequence of the nucleotides that make up the gene. The beginning point along this strand is marked by a particular sequence of DNA nucleotides (TAC). RNA nucleotides from the cytoplasm enter the nucleus and bond in a complementary fashion to the nucleotides of the DNA strand bearing the marker sequence. The RNA nucleotides then bond together forming the linear mRNA molecule. This sequence of events is **transcription**, namely the making of mRNA from a DNA **template**. Recall that thymine nucleotides do not exist in RNA; the very similar base **Uracil** (U) replaces Thymine.

Once constructed, mRNA leaves the nucleus and becomes associated with a ribosome where the protein will actually be synthesized. It is important to understand the nature of the chemical message of mRNA. There are only four bases in RNA (A, U, C, or G), and therefore, only four

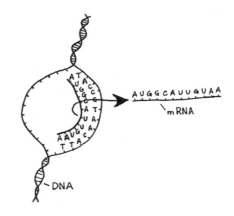

Figure C-3. Transcription. *This process allows the genetic message (the code) to be copied ("transcribed") into a form that can leave the nucleus. This copy is known as mRNA. mRNA contains a base sequence (a set of codons) that is complementary to the gene. Uracil replaces thymine in RNA. Note that the start codon is AUG, which codes for methionine.*

different nucleotides. There are 20 different amino acids that could be required to form the protein. Obviously, these nucleotides cannot each stand for an amino acid. If they were considered in pairs, there are only 16 combinations, which still isn't enough. They actually function in groups of three, or **triplets** to provide 64 combinations. Each triplet of mRNA nucleotides is called a **codon**. DNA is said to contain a **code** for protein synthesis; mRNA's version of this code is a sequence of codons. Most amino acids have more than one triplet codon to govern their inclusion into the forming protein.

    Translation, the second phase of protein synthesis, occurs at ribosomes. This part of the process can be summarized into steps as follows:

1. The mRNA arrives at a ribosome where **ribosomal RNA (rRNA)**, part of the ribosome structure, helps align it into the correct position on the ribosome. The mRNA consists of a series of codons complementary to the gene in the parent DNA strand. The first codon is AUG (the complement of TAC), which codes for the amino acid, methionine. AUG is the "**start codon**"; methionine is the first amino acid in any protein. This step is called **initiation**.

2. Initiation is followed by **elongation**, which involves putting the amino acids into the correct sequence to build the required protein. Elongation requires a different type of RNA called **transfer RNA (tRNA)**. The shape and structure of tRNA molecules is very precise. Each one has a bonding site for a specific amino acid, is the exact same length, has an **anticodon** on the end opposite to the amino acid, and a three dimensional configuration that allows it to be held at a ribosome in a position that allows the formation of peptide bonds between the amino acids. Anticodons are complementary to the mRNA codons. It follows that there are sixty-four anticodons and, therefore, sixty-four different tRNA molecules (though some of them bear the same amino acids as others). The role of a tRNA molecule is to transport a specific amino acid to the ribosome.

3. At the ribosomes, the exposed codons dictate which tRNA molecules are required in which sequence (due to complementary base pairing). As the correct tRNA molecule is aligned by its anticodon pairing with the codon, its amino acid is also aligned at the opposite end. The sequential alignment of the amino acids promotes the formation of **peptide bonds** between them. It starts as methionine bonds to the next amino acid (making a dipeptide). In this fashion, the growing amino acid chain is bonded to the next amino acid in sequence. Once the amino acid of a t-RNA molecule is bonded into sequence, the t-RNA molecule (freed of its amino acid) returns to the cytoplasm to gain another amino acid so that it can be used again.

4. This process continues until a **terminator** (stop) **codon** is reached. The tRNA that is complementary to a terminator codon does not transport an amino acid; thus the sequence is ended. Of the 64 codons, three are terminator codons. This last step is appropriately called **termination**.

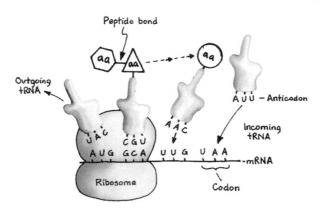

Figure C-4. Translation. *The subunits of a ribosome assemble and pass along the mRNA strand reading it codon by codon. Each codon is associated with a tRNA molecule that has a complementary base sequence known as an anticodon. These tRNA molecules bring in the required sequence of amino acids, which are assembled one at a time to make the protein. The peptide chain is passed on to the incoming amino acid as it grows until the stop codon is reached. The final tRNA molecule lacks an amino acid, ending the amino acid sequence and the finished protein is released.*

    To summarize, protein synthesis occurs in several steps. The synthesis of mRNA from the DNA template occurs first. This is transcription - where the DNA code is *transcribed* into the codons of mRNA. The second part occurs at the ribosomes with the arrival of a strand of mRNA. This is where the codon sequence is *translated* into an amino acid sequence forming a polypeptide (protein). This second part of protein synthesis first has to be **initiated** by the start codon, then the amino acid chain is **elongated** one amino acid at a time, and finally it is **terminated** by a stop codon.

    The **genetic code** (see Appendix II) is a list of the 64 possible codons and the amino

---

acids they code for. The genetic code is universal, meaning it holds true for all organisms. With it, one can determine the amino acid sequence for a given DNA strand described by a specific set of bases. For example, the portion of a DNA strand containing the base sequence CCTGACCTG codes for the amino acid sequence of Glycine-Leucine-Aspartate. These amino acids, once assembled become part of a protein (enzyme, antibody etc.). It is also possible to recognize the effects of altering the base sequence in DNA. Obviously changing the DNA could have serious implications for protein production. This is the topic of mutations.

C.2 CONCEPT CHECK-UP QUESTIONS:
1. How is mRNA formed? What is its role?
2. Name the bases on the strand of mRNA produced from the following segment of DNA: TCTATGCTC. What amino acids does this strand represent?
3. What is the significance of methionine?
4. Explain "termination".

MUTATIONS

The term "**mutation**" has become sort of a "catch-all" for biological errors. There are really two kinds of mutations, **gene (or point) mutations** and **chromosome mutations**. A gene mutation occurs when an error is made during one of the processes that involve base pairing between nucleotides. There is a tendency to consider that gene mutations only occur during replication because an error here would exist in every subsequent daughter cell produced from the parent cell. If it were in a gamete, then the error would be in every cell of the resulting organism! If a gene mutation were to occur during transcription or translation, it would be unnoticed for it would only affect the formation of a single protein in a single cell of an organism. Chromosome mutations, in contrast, involve chromosomes being broken or "mislocated" during cell division as in the case of **trisomy 21**.

All cell processes should occur normally; they have to be influenced by some external factor to do otherwise. A factor that causes a mutation is called a **mutagen**. The two most common mutagens are radiation and chemicals.

This unit considers three types of gene mutations.
1. *Deletion* - occurs during base pairing when, somehow, a single nucleotide gets left out. This means the gene in a new strand of DNA will be one nucleotide short. Obviously, any time this gene undergoes transcription the set of triplet codons would be incorrect from the point of the deletion on.
2. *Addition* - similar to deletion, only this time a nucleotide is inserted along an otherwise normal DNA strand. In this case, the gene is one nucleotide longer than it should be, and, again, the set of codons is incorrect from the site of addition on.
3. *Substitution* – in this case the entire base sequence is correct except in one location where one nucleotide replaces another. The mRNA strand produced will contain all the right codons except for the one that contains the substituted nucleotide. This change *may* affect the incorporation of the correct amino acid into the protein.

A protein that is produced following the deletion or addition of a nucleotide is likely nonfunctional. Cells with this type of mutation probably die, particularly if the protein that was supposed to be made is an essential enzyme.

In the event of substitution, where mRNA with a slight error in its codons can be produced, the outcome depends on the exact error. Consider changing the codon from CUA to CUG. There is no effect. The codon still codes for leucine. The genetic code is termed **degenerative** for this reason - meaning that the severity (or significance) of changes to the codon degenerate (lessen) as the triplet is read (L to R). To stay with the same example, if a substitution were to convert CUA into CAA, the effect would be to incorporate an entirely different amino acid (glutamine instead of leucine) into the protein. A protein would still be produced, but it may never function properly because of the wrong amino acid.

Proteins, once formed (regardless of their condition), are intended for some purpose (such as forming structures or functioning as hormones, neurotransmitters, antibodies, enzymes, etc). Many are used in the cell; others are packaged up and exported. Mutant proteins may reach their appropriate destinations, but will fail to function properly.

RECOMBINANT DNA

The origins of DNA technology date back to research in the first half of the 20th century. It wasn't until the 1970's that the branch of biological study known as **recombinant DNA** technology began to take off. It has the potential to revolutionize human lives on earth.

Simply put, recombinant DNA technology allows scientists to insert a segment of DNA (a gene) from one organism into a chromosome of another. When the host organisms are bacteria, these genes can be mass-produced easily, quickly, and at a very low cost. *E. coli*, a relatively common and harmless human bacterium, is easy to maintain **in vitro** (in the lab) and it will undergo binary fission about every 20 minutes under good conditions.

These are some significant advantages to conducting these techniques:

1. When a desirable gene (one that codes for a desired protein) is inserted into a bacterium, it can be used to mass-produce the protein it codes for. Examples of proteins that are currently mass-produced in this way include some hormones (e.g., insulin and growth hormone), various antibodies, and proteins used in cancer treatment (e.g., interferon).
2. A similar technique transforms bacteria thus altering their original metabolism. This technique was used to produce a strain of oil-metabolizing bacteria, which may have implications for oil spill clean-ups.
3. Recombinant DNA is being used to produce hybrid species of some food crops that are disease resistant or more tolerant of extreme environmental conditions. This is leading to a greater production of food worldwide.
4. Currently, recombinant DNA techniques are being used to minimize the effects of genetic errors in humans. In this technique, cells from a person suffering a genetic abnormality can be removed, "repaired" and reinserted. Through this type of **genetic engineering**, affected individuals can sometimes gain some of the benefits of the normal functioning cells.
5. The laboratory manipulation of genes has greatly increased the ability to produce genetic **clones**, which are useful because organisms with identical genetic make-up can be used in studies of differential gene expression. Cloned organisms may also play a huge role in food and resource production.

DNA research is not being done without objection. There are some ethical issues and public paranoia associated with it. Recombinant DNA techniques give researchers the ability to design organisms, literally create new species of organisms on earth. Additionally, there are some public advocates against the consumption of genetically altered or enhanced food. Other advocates speak out about the rights of animals and are appalled by the production of laboratory clones.

C.3 CONCEPT CHECK-UP QUESTIONS:
1. Explain why some gene mutations may remain unnoticed throughout one's life.
2. Suppose a DNA gene mutation were to cause adenine to be incorporated instead of guanine during transcription. How would this alter the intended polypeptide sequence for this segment of DNA: **AGGTTCTGA**?
3. Define recombinant DNA.
4. List two ways that recombinant DNA techniques could affect our day-to-day lives.

CHECK YOUR UNDERSTANDING

MULTIPLE CHOICE:

1. Nucleic acids are composed of chains of
   A. genes.
   B. purines.
   C. pyrimidines.
   D. nucleotides.

2. DNA has equal amounts of A and T because
   A. these molecules bond together.
   B. DNA is a double stranded helix.
   C. these molecules are not the same size.
   D. one is a purine and one is a pyrimidine.

3. In a DNA molecule, each cross-member (or rung) has two
   A. sugars.
   B. purines.
   C. phosphates.
   D. nitrogenous bases.

4. DNA of unrelated species differs in the
   A. type of sugar.
   B. number of strands.
   C. sequence of bases.
   D. order of phosphates.

5. DNA is the "master molecule" because it
   A. is a double helix.
   B. is located in the nucleus.
   C. ultimately controls all cell activity.
   D. stores energy as high energy bonds.

6. Replication refers to the synthesis of
   A. complementary strands of RNA for mitosis.
   B. complementary strands of DNA for mitosis.
   C. proteins based on the sequence of bases in RNA.
   D. proteins based on the sequence of bases in DNA.

7. The two molecules resulting from the replication of DNA differ
   A. only if mutations occur.
   B. in terms of function only.
   C. because one is old and one is new.
   D. because one has ribose and the other has deoxyribose.

8. Which is mass-produced directly by bacteria altered by recombinant DNA?
   A. Insulin.
   B. Glycogen
   C. Cholesterol.
   D. Testosterone.

9. A polymer of nucleic acids found in the cytoplasm of a living animal cell would
   A. be helical.
   B. contain ribose.
   C. contain thymine.
   D. be double stranded.

10. Genetic engineering **CANNOT**
    A. change a person's genetics.
    B. develop hybrid species of crops.
    C. produce clones for scientific investigations.
    D. mass produce molecules for medical treatments.

11. For any given strand of DNA
    A. $[A] = [C], [T] = [G]$
    B. $[A] = [G], [T] = [C]$
    C. $[A] + [T] = [G] + [C]$
    D. $[A] + [G] = [T] + [C]$

12. What percentage of G does a chromosome with 27% A have?
    A. 23%
    B. 27%
    C. 46%
    D. 54%

13. Which is **FALSE** about replication?
    A. New molecules have one old strand.
    B. DNA unzips as H-bonds break.
    C. Each base gets paired with another identical one.
    D DNA helicase and DNA polymerase help the process.

14. Genetic information flows from
    A. Protein to RNA to DNA
    B. DNA to RNA to protein
    C. DNA to protein to RNA
    D. RNA to DNA to protein

15. How many bases (nucleotides) code for one amino acid?
    A. 1
    B. 3
    C. 20
    D. 64

16. During transcription, the information in
    A. RNA is converted into DNA.
    B. DNA is converted into RNA.
    C. protein is converted into RNA.
    D. RNA is converted into protein.

17. What are ribosomes composed of?
    A. rRNA only.
    B. proteins only.
    C. rRNA and protein.
    D. mRNA, rRNA, and protein.

18. Using the Genetic Code requires the
    A. amino acid sequence.
    B. code sequence from DNA.
    C. codon sequence from mRNA.
    D. anticodon sequence from tRNA.

19. Which anticodon corresponds to ATC?
    A. TUC
    B. AUG
    C. AUC
    D. ATC

20. The complementary DNA sequence to AAGCTT is
    A. UUCGAA
    B. TTCGAA
    C. AACGTT
    D. TTCGUU

21. The Genetic Code does **NOT**
    A. have 64 codons.
    B. include base triplets.
    C. vary among species.
    D. contain start and stop codons.

22. Initiation, elongation, and termination are three stages in
    A. transcription.
    B. DNA replication.
    C. polypeptide synthesis.
    D. codon-anticodon formation.

23. What signals the end of translation?
    A. A terminator codon.
    B. The cell runs out of amino acids.
    C. The ribosomes run off the end of mRNA.
    D. Enzymes signal that the polypeptide is long enough.

24. Protein synthesis involves
    A. Transcription in the nucleus before translation at ribosomes.
    B. Transcription at the ribosomes before translation in the nucleus.
    C. Transcription in the nucleus after translation at the ribosomes.
    D. Translation in the nucleus after transcription at the ribosomes.

25. Which is **FALSE** about mRNA?
    A. It is made at the ribosomes.
    B. It is made from a DNA template.
    C. It can be found in the cytoplasm.
    D. It contains uracil in place of thymine.

26. During transcription
    A. Deoxyribose turns into ribose.
    B. Uracil matches up with adenine.
    C. Uracil matches up with thymine.
    D. Deoxyribose is used instead of ribose.

27. Which sequence **BEST** outlines protein synthesis?
    A. mRNA, DNA, tRNA, amino acids
    B. DNA, mRNA, tRNA, amino acids
    C. DNA, tRNA, amino acids, mRNA
    D. DNA, amino acids, mRNA, tRNA

28. What type of gene mutation converts TACGGCATG to TAGGCATG?
    A. Deletion.
    B. Addition.
    C. Substitution.
    D. Chromosomal.

29. Anticodons differ from codons as they
    A. attach to ribosomes; codons don't.
    B. contain thymine; codons contain uracil.
    C. are part of tRNA; codons are part of mRNA.
    D. relate to amino acids; codons relate to nucleotides.

30. Incorrect base pairing **MOST** likely only causes a problem during
    A. replication of DNA.
    B. transcription forming mRNA.
    C. translation if rRNA misaligns mRNA.
    D. translation when tRNA brings in amino acids.

31. What sequence of these steps occurs during translation?
    1. Peptide bonds form.
    2. Anticodons match codons.
    3. tRNA molecules are freed into the cytoplasm.
    A. 3, 1, 2
    B. 2, 3, 1
    C. 2, 1, 3
    D. 3, 2, 1

32. Which strand could form by substitution during the replication of CATUAUCCC?
    A. ATUAUCCC
    B. CTUAUCCC
    C. CATUAUCGC
    D. CATTUAUCCC

33. Which type of gene mutations can have the LEAST impact on protein synthesis?
    A. Deletion.
    B. Addition.
    C. Substitution.
    D. Any. They are all about the same.

The next NINE questions require the Genetic Code.

34. If UGG was mutated to UGA during transcription,
    A. nothing would change.
    B. the protein would be too short.
    C. the anticodon mutates to ACU.
    D. threonine would get incorporated into the protein.

35. What mRNA strand codes for the amino acid sequence below?

    cys – leu – phe – ala – glu

    A. UGC CUA CCU GCU GAA
    B. UGG CUU UUU GCA GUC
    C. UGU CUC UUG GCC GAC
    D. UGU UUG UUU GCG GAG

36. In which triplet codon would a single substitution in the third position have the greatest mutational effect?
    A. GUU
    B. AUU
    C. CGU
    D. AUG

37. Suppose a DNA sequence mutated from
    AGAGAGAGAGAGAGAGAG
    to
    AGAAGAGAGATCGAGAG
    What amino acid sequence will be generated based on this mutated DNA?
    A. ser-ser-leu
    B. ser-leu-ser-leu-ser-leu
    C. glu-arg-glu-leu-leu-leu
    D. leu-phe-arg-glu-glu-glu

38. The DNA that codes for the polypeptide **Phe-Leu-Ile-Val** is
    A. TTG-CTA-CAG-TAG
    B. AAA-AAT-ATA-ACA
    C. AUG-CTG-CAG-TAT
    D. AAA-GAA-TAA-CAA

39. What amino acid strand is formed from the DNA strand?

    ATA–CGA–CAA–GCC

    A. tyr–ala–val–arg
    B. tyr–arg–leu–ala
    C. met–ala–leu–ala
    D. met–arg–glu–cys

40. The amino acids phenylalanine, alanine, and lysine exist in sequence in a protein. Which DNA sequence would cause proline to substitute for alanine?
    A. AAA-GGG-TTT
    B. AAA-CGG-TTT
    C. AAA-CCG-TTT
    D. AAA-CCC-TTT

41. What amino acid sequence results from the following mRNA codon sequence?

    AUG-UCU-UCG-UUA-UCC-UUG

    A. met-ser-leu-ser-leu-ser
    B. met-ser-ser-leu-ser-leu
    C. met-glu-arg-arg-gln-leu
    D. met-arg-glu-arg-glu-arg

42. Which statement does **NOT** illustrate the degenerative nature of the Genetic Code?
    A. There are three codons for isoleucine.
    B. Any base following a CC_ sequence refers to proline.
    C. Even if C replaces A, the triplet AGG still positions arginine.
    D. If G replaces the last A in UAA, it is still a stop codon.

Use the following diagram to answer the next question.

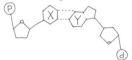

43. Which of the following could be true?
    A. X = uracil; Y = adenine.
    B. X = adenine; Y = thymine.
    C. X = guanine; Y = cytosine.
    D. X = thymine; Y = guanine.

WRITTEN ANSWERS:

1. Describe the structure of DNA. In what ways is it distinct from RNA?

2. List the three functions of DNA and describe the biological significance of each one

3. Design a data table to contrast "code", "codon", and "anticodon" in terms of location, structure and function. Complete your data table.

4. What are the three types of RNA and what are their specific functions?

5. What are the roles of start and stop codons?

6. A segment DNA was analyzed and calculated to contain 26 Guanine molecules and 34 adenine molecules. How many nucleotides does this segment of DNA contain all together? Explain how you determine your answer.

7. **BRIEFLY** describe each of the following and describe its effect on the formation of a protein.
    a. base deletion
    b. base addition
    c. base substitution

8. Consider the process of **replication**.
    a. Sketch this process.
    b. Include base sequences in your sketch.
    c. Explain why the resulting DNA molecules are identical to each other and the parent strand.

9. Consider the process of **transcription**.
    a. Sketch this process.
    b. Put a random base sequence on your DNA molecule.
    c. Draw in the product of this process. Label it and identify its base sequence.
    d. Explain what happens to the product of this process.

10. Consider the process of **translation**.
    a. Sketch this process.
    b. Include appropriate base sequences in your diagram.
    c. What is the product of this process? Explain what happens to this product.

11. Consider the three processes (in Questions 8, 9, and 10). In which one would a mutation have the most severe consequences? Explain.

UNIT D - MEMBRANE STRUCTURE AND FUNCTION

All cells have cell membranes that control and regulate the passage of materials across their boundaries. The chemical nature of these membranes differentially restricts and promotes the movement of particles through them. Similarly, most organelles have membranes distinguishing their contents from the cytoplasm. The only membrane that prokaryotic cells have is a cell membrane. They lack all internal membranes and, therefore, all membrane-bound organelles.

It is important for cells and membrane-bound organelles to be able to regulate the inward and outward passage of materials. This helps them control their contents. While they are at the mercy of the **Law of Diffusion**, and will gain or lose ions or water because of environmental concentrations, they are dynamic in their ability to participate in this movement. Membranes have specialized proteins embedded in them that serve as transportation routes for certain substances into or out of cells. Sometimes this passage requires energy, other times it does not.

Many cells also require the ability to move large substances through their membranes – substances that are too large to pass through the biochemical structure, itself. In cases like these, the membranes have to be "broken and reformed". Knowledge of membrane structure and the nature of the biochemicals present in membranes is essential to understanding the function of membranes.

THE NATURE OF MEMBRANES

The primary substance of all membranes is a **phospholipid bilayer**. The arrangement of the **phospholipids** provides the outer edges of the membrane with a **polar surface**. For this reason, cells exist with polar molecules such as environmental water on the outside and their watery cytoplasm on the inside. A good example of this is blood where blood cells are transported by plasma that is over 90% water.

Not all phospholipids are the same (due to the different lengths and types of fatty acids they are made out of), and therefore the specific nature of the interior of the phospholipid bilayer is somewhat variable. Regardless, all these interior regions are non-polar and have an oily consistency. It is because of this that water, unlike fatty acids and other lipid soluble molecules, cannot pass freely though membranes.

Proteins embedded in the phospholipid bilayer provide the membrane with its specific regulatory capabilities and functions. Some proteins, called **peripheral proteins** are on the outer surface and serve the cell as **receptor sites** by interacting with environmental molecules, like neurotransmitters or hormones, and impact the cell. Other proteins, called **integral** or **transmembrane** proteins, provide the mechanisms for lipid-insoluble materials to cross the membrane. Water, for example, passes through **protein channels**, which are polar routes through the 3-dimensional structure of some proteins. Other lipid-insoluble substances such as ions, amino acids and glucose actually bond to particular proteins and are propelled through the membrane, sometimes requiring the expenditure of **ATP** energy. Still other membrane proteins have enzymatic functions or can serve a cell as cytoskeletal anchors. The variety of protein structures found in membranes contributes immensely to the specific **permeability** of the membrane.

The **fluid-mosaic model** has been developed to illustrate the molecular components of the cell membrane and is used to explain how these components function together. The proteins form a mosaic pattern within the phospholipids. The phospholipids provide a fluid nature to the membrane. Animal cell membranes also have cholesterol molecules embedded between the phospholipids. Cholesterol is produced by the liver and is transported by plasma to cells throughout an animal's body. Cholesterol fills the spaces between the phospholipids and helps maintain the integrity of the membrane structure.

[handwritten notes in right margin:] carbohydrates cholesterol glycolipids filaments

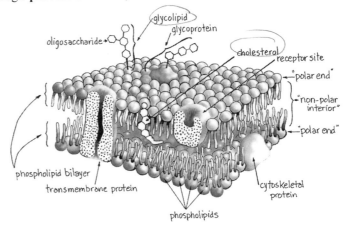

Figure D-1. Fluid Mosaic Model. *Membranes are made of double layers of phospholipids arranged so their polar ends are to the outside making them hydrophilic. Embedded in this structure is a mosaic of different types of proteins enabling the membrane to have certain roles (transport, reception etc.). The outer surfaces of the membranes of animal cells have oligosaccharides attached to proteins and phospholipids forming glycoproteins and glycolipids (respectively).*

The outer surface of an animal cell membrane has two kinds of oligosaccharide attachments. Those attached to phospholipids are called **glycolipids**; those attached to proteins are called **glycoproteins**. These attachments function as markers for the identification of cell types. A blood **antigen** is a carbohydrate sequence attached to a protein in the membrane of a red blood cell. Different antigens (glycoproteins) like these contribute to the differences between blood types. Plant cell membranes and the membranes of organelles lack these carbohydrate chains.

The term **selectively permeable** is used to describe membrane function. This term implies that membranes can select (or deselect) substances that are able to move through them. Historically, the term that was used was "semi-permeable" to imply that some things could cross a membrane where others could not. In keeping with the fluid mosaic model of membrane structure, this term has been dropped in favor of "selectively permeable", which acknowledges the more active role membranes, and particularly their proteins, play in regulating the nature of the contents of cells. **SPM** is the acronym for **s**electively **p**ermeable **m**embrane.

Clearly, the differences in the biochemical composition of membranes account for the differences in their permeability. All membranes have the same fundamental structure but tremendous variations exist due to the variety of phospholipids, proteins and oligosaccharides that form them.

D.1 CONCEPT CHECK-UP QUESTIONS:
1. Name three types of biochemicals located in membranes and describe a function of each one.
2. Explain why cells mix freely with water.
3. Why is "selectively permeable" a better term to describe membranes than "semi-permeable"?
4. Will all the cell membranes in a particular animal have the same permeability? Explain.

TRANSPORT ACROSS MEMBRANES

DIFFUSION

The random movement of particles due to concentration differences is called **diffusion**. This is a **passive** process meaning energy (ATP) is not required. In fact, a membrane isn't even required! A smell will diffuse through the air. Similarly, some substances diffuse through membranes as if the membrane wasn't even there. Molecules will continue to diffuse until they are evenly distributed (i.e., eventually, a whole room will harbor the same smell). This natural phenomenon is called the **Law of Diffusion**. A number of factors affect the rate of diffusion:

1. *Concentration*. **Solutions** consist of two parts: **solvent** (often water) and **solute** (particles dissolved in the solvent). Both the cytoplasm of cells and the extracellular fluid around cells are solutions. A difference in solute concentration between two areas (**concentration gradient**) causes diffusion. The greater the difference is, the faster the diffusion. For example, if the concentration of oxygen outside a cell (in the extracellular fluid) is greater than it is inside the cytoplasm of the cell, the oxygen will diffuse into the cell. Oxygen is the solute in this example.
2. *Temperature*. Increasing the temperature of a solution allows the particles to move more (increases their kinetic energy). This increased motion increases diffusion.
3. *Size and Shape*. Smaller, more streamlined substances will diffuse more rapidly because they encounter fewer collisions with other substances. This factor is a critical one when considering the diffusion across a membrane. Large molecules (starches, proteins etc.) do not diffuse across membranes.
4. *Ionic Charge*. By virtue of their own electrical nature, ions are attracted or repelled into particular directions assisting or hindering diffusion depending on the situation.
5. *Viscosity*. The lower the viscosity ("fluid density") of the solution, the more quickly molecules can move through it. By way of example, the viscosity of water is much less than that of syrup. Diffusion through water occurs more quickly than through syrup for this reason.
6. *Movement of the Medium*. Currents will aid diffusion. In the air, it is easy to consider how wind may help disperse molecules. In cells, a phenomenon called **cytoplasmic streaming** (constant movement of the cytoplasm) does the same thing.

In addition to these factors, there are two general conditions that affect a particle's ability to pass freely through a membrane: **solubility** and **polarity**. Small lipid soluble molecules will move through the phospholipid bilayer quite readily, as do small non-polar gas molecules like CO_2 and O_2. Water also diffuses, but because of its polarity, it will not pass through the non-polar interior of the bilayer. Instead, water passes through specialized protein channels.

OSMOSIS

The diffusion of water is known as **osmosis**. Like other molecules, water moves to even out concentrations. Water moves to areas with greater solute concentrations (= *up* a solute concentration gradient). This dilutes the solutes, effectively equalizing the concentration throughout. Solutions with equal solute concentrations are termed **isotonic**.

Osmosis is very significant in membranous systems. Membranes, being selectively permeable, often restrict the movement of some molecules - particularly large ones. As a result, water molecules travel rather freely through protein channels to maintain suitable solute concentrations. **Osmotic pressure** is a measure of the number of collisions water molecules make against the membrane surface. The greater the difference in solute concentration is, the greater the number of collisions, which means a higher osmotic pressure. **Turgor pressure (turgidity)** is the force the cytoplasm exerts against the inside of the cell membrane. Significant changes in turgidity can affect cell function.

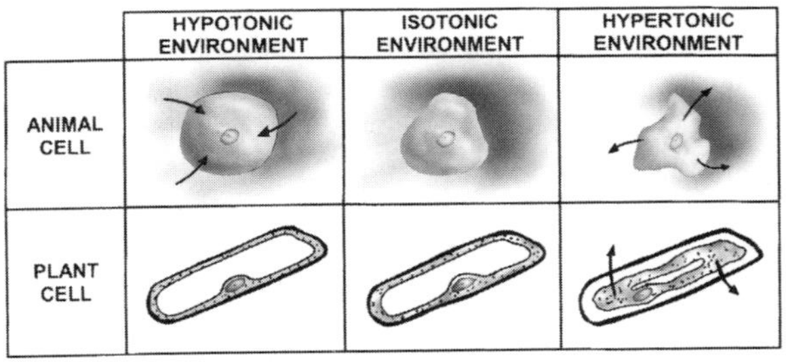

When a cell is exposed to an environment with a greater solute concentration (called **hypertonic**), water will leave the cell by osmosis, thus diluting the solute in its environment. At the same time it is concentrating the solutes in its cytoplasm. This process is called **plasmolysis**. It results in decreased turgidity and the shriveling (or shrinking) of animal cells and the **protoplasm** of plant cells. A plant cell wall is sturdy enough that it resists movement and, when viewed with a microscope, the cell membrane can be seen "wilting" away from the cell wall. Some cells, such as red blood cells undergo **crenation** during plasmolysis, which means they take on a notched appearance. Prolonged plasmolysis causes cell death.

Figure D-2. Osmosis. *Cells gain and lose water depending on the relative difference in concentration between their cytoplasm and their environment. In a hypotonic environment, cells will gain water, which increases their turgidity. This causes swelling in animal cells. No significant change is noticeable in plant cells due to the rigidity of the cell wall. In hypertonic environments, both cell types will lose water. The animal cell and the protoplasm of the plant cell both shrivel. The cell wall remains intact.*

The opposite scenario also exists. When a cell is subjected to an environment with a lesser concentration of solutes, such as distilled water, water will enter the cell by osmosis (again resulting in the dilution of the greater solute concentration). This is **deplasmolysis**. For animal cells this causes a gain in turgidity and swelling. The membranes of red blood cells placed in distilled water or some alcohols can rupture due to the rapid osmosis and increased turgidity. This phenomenon is known as **hemolysis**. In the case of plant cells, the inward osmosis increases the turgidity, but the cell wall resists movement.

D.2 CONCEPT CHECK-UP QUESTIONS:
1. How are diffusion and osmosis similar? Different?
2. Why might particles be unable to enter or leave a cell even though a significant difference in concentration exists?
3. Given that cell membranes contain protein channels, what can restrict the movement of water in or out of the cell?
4. Why does watering a wilted plant make it "perk up"?

PROTEIN ASSISTED TRANSPORT

Some integral proteins embedded in the phospholipid bilayer of a cell membrane are involved in the transport of substances other than water. These transmembrane proteins have three-dimensional configurations with surface regions that are sensitive to the

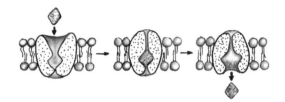

Figure D-3. Protein Assisted Transport. *Proteins that extend all the way through a membrane often function as transport proteins by propelling ions or small molecules through the membrane. If the direction of movement is down a concentration gradient, then the process is termed facilitated transport. If the direction is the other way, ATP is required and the process is known as active transport.*

presence of specific solute particles. The specificity of these protein carriers contributes to the selective nature of the membrane. When the solute particles attach to the protein, they cause a change in its shape, which propels the solute particles through the membrane. There are two variations of this:

1. ***Facilitated transport*** - where the substances are transported according to the Law of Diffusion (down a concentration gradient). No energy is required for this transport.

2. ***Active transport*** - where the substances are transported against the Law of Diffusion, i.e., up a concentration gradient. For example, the **thyroid gland** accumulates the iodine needed to manufacture the hormone **thyroxin**. The iodine concentration can be as much as 25 times greater in the tissues of the thyroid than in blood plasma. Another example is the **Na/K pump**, which restores the electrical order in a neuron after an impulse has traveled along it. Active transport requires energy from the hydrolysis of ATP.

TRANSPORT INVOLVING VESICLES

The largest substances that have to enter or leave a cell do through the use of membranous vesicles. When cytoplasmic vesicles containing wastes or products for secretion come in contact with a cell membrane, they fuse onto it and open its surface, thus spilling their contents to the outside. This is **exocytosis**. **Endocytosis** is a similar sort of process only in reverse.

The best-documented example of endocytosis is an ameba ingesting food by **phagocytosis**. The result of this ingestion is the formation of a **food vacuole**. This vacuole fuses with a lysosome containing hydrolytic enzymes, which digest the vacuole's contents. **Pinocytosis** is a reduced version of phagocytosis where smaller objects, or fluids are ingested and a vesicle (as opposed to a food vacuole) is immediately formed. The expressions "cell eating" and "cell drinking", respectively, are often used to clarify the difference between phagocytosis and pinocytosis. All these processes require the expenditure of ATP energy.

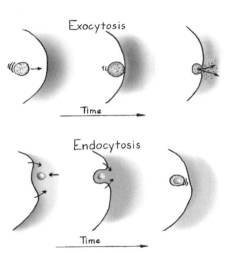

Figure D-4. Exocytosis and Endocytosis. *Vesicles transport substances that are otherwise too large to fit through the membrane. Cytoplasmic vesicles fuse with the membrane to release their contents to the outside. This is exocytosis. Substances are taken into a cell by endocytosis forming vesicles, such as food vacuoles.*

REQUIREMENTS OF CELLS

Cells exist individually, or in clusters forming tissues, organs, and eventually organisms. Cell size is a very important factor for their survival. Cells are the size they are because that is the size they have to be in order to function effectively. As a cell goes about its regular activities, the substances needed and the wastes produced must pass through its cell membrane. When the composition of a cell's environment is altered, osmosis causes the cell to change size to accommodate the change in the environment. For example, putting an animal cell in distilled water will cause it to swell. The cell's volume will increase, but its surface area will remain the same.

When the volume of cells changes, they suffer the loss of their normal **surface area/volume ratio** [the volume increases faster than the surface area (cubed function vs. squared function)]. The surface area becomes the limiting factor in the cell's ability to survive. By way of example, a swollen cell may produce more wastes than it can get rid of, or it can't consume enough materials to function properly at the increased size. Apart from dying there are three possible solutions to this change: altering shape, altering function, or dividing. Unfortunately, specialized cells of multicellular organisms can't do any of these.

Changing a cell's shape and function are not viable alternatives. Cell shape and function are related - cells are the shapes they are because their shape allows them to function and stay alive. For example, nerve cells are often long and thin for transmitting impulses, absorptive cells often have convoluted margins to maximize surface area, and so on. Cell division increases the SA / V ratio, but often cells are too specialized to divide. If changes in environmental conditions are too extreme and too long lasting, cells die.

Cell membranes allow cells to obtain all their raw materials from their environment, and get rid of various products and by-products. The relationship between cell volume and surface area is critical to survival.

D.3 CONCEPT CHECK-UP QUESTIONS:
1. Contrast facilitated and active transport.
2. Are cell membranes permeable to Na^{1+}? By what mechanism does Na^{1+} cross cell membranes?
3. Since endocytosis produces vesicles from cell membranes, suggest why cells that conduct endocytosis do not get smaller.
4. Why does doubling the surface area of a cell not double its volume? Relate this concept to cell metabolism.

CHECK YOUR UNDERSTANDING

MULTIPLE CHOICE:

1. Particles are expelled from cells by
   A. exocytosis.
   B. endocytosis.
   C. phagocytosis.
   D. reverse osmosis.

2. Energy is required for
   A. Osmosis.
   B. Diffusion.
   C. Active transport.
   D. Facilitated transport.

3. What must exist for diffusion to occur?
   A. A solution.
   B. A living cell.
   C. A concentration difference.
   D. A selectively permeable membrane.

4. Wilted plant tissue placed in cold water will become stiffer and harder because the
   A. turgidity decreases.
   B. water passes into the cells.
   C. cellulose walls get thicker.
   D. cell's starch becomes glucose.

5. Pinocytosis and phagocytosis are accomplished by the
   A. nucleus.
   B. mitochondria.
   C. cell membrane.
   D. endoplasmic reticulum.

6. Proteins do not pass freely through membranes because they
   A. contain nitrogen.
   B. would get denatured.
   C. are very large molecules.
   D. are components of membranes.

The next question refers to this sketch.

7. What sequence of events is depicted?
   A. Cell eating, then cell drinking.
   B. Phagocytosis, digestion, then exocytosis.
   C. Endocytosis, translation, then exocytosis.
   D. Active transport, digestion, then exocytosis.

8. When glucose crosses a membrane, it
   A. uses a carrier molecule.
   B. enters through a vesicle.
   C. passes between phospholipids.
   D. goes through protein channels.

9. Animal cells take in water when in
   A. osmotic solutions.
   B. isotonic solutions.
   C. hypotonic solutions.
   D. hypertonic solutions.

10. Which shape of cells would have the **LOWEST** SA/V ratio?
    A. Flat.
    B. Square.
    C. Irregular.
    D. Spherical.

11. Which substance needs a vesicle to cross a membrane?
    A. Starch.
    B. Glycerol.
    C. Amino acids.
    D. Monosaccharides.

12. Which normally crosses a membrane by active transport?
    A. Water.
    B. Oxygen.
    C. Sodium ions.
    D. Carbon dioxide.

13. What component of a cell membrane functions as a carrier molecule?
    A. Protein.
    B. Glycolipid.
    C. Glycoprotein.
    D. Phospholipid.

14. The Law of Diffusion is **NOT** obeyed by
    A. Osmosis.
    B. Exocytosis.
    C. Active transport.
    D. Facilitated transport.

15. Which is hypertonic to 0.9% NaCl?
    A. 0.7% NaCl
    B. 0.8% NaCl
    C. 0.9% NaCl
    D. 1.0% NaCl

16. When a substance moves from an area of low to high concentration,
    A. energy is needed.
    B. cells begin to rupture.
    C. diffusion has occurred.
    D. the SA/V ratio changes.

17. The difference in solute concentration between two areas/regions is called
    A. hypotonic.
    B. hypertonic.
    C. absolute difference.
    D. concentration gradient.

18. Lipids pass through membranes because
    A. they get enveloped in vesicles.
    B. of the availability of ATP energy.
    C. membranes are largely non-polar.
    D. they are assisted by protein carriers

19. Which term refers to cell rupturing?
    A. Lysis.
    B. Turgor.
    C. Crenation.
    D. Plasmolysis.

20. If the solute concentration of solution A is greater than solution B, then A is
    A. isotonic to B.
    B. osmotic to B.
    C. hypotonic to B.
    D. hypertonic to B.

21. Endocytosis requires
    A. diffusion.
    B. vesicle formation.
    C. transport proteins.
    D. secretory vesicles.

22. When red blood cells are put into very salty solutions,
    A. they burst.
    B. they crenate.
    C. water enters the cells.
    D. proteins pump salt into the cells.

23. The Law of Diffusion confirms that substances move along a concentration gradient in a
    A. upward direction using ATP.
    B. downward direction using ATP.
    C. upward direction not using ATP.
    D. downward direction not using ATP.

24. If the mitochondria in a cell were **NOT** functioning properly then
    A. the cell couldn't get oxygen.
    B. phagocytosis could not occur.
    C. osmotic pressure would decrease.
    D. the cell couldn't absorb fatty acids.

25. Which describes osmotic pressure?
    A. Collisions of water on a membrane.
    B. Volume of water crossing a membrane.
    C. Size of water compared to solute particles.
    D. Movement of water to higher solute concentrations.

The next question refers to this sketch.

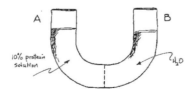

26. During this experiment, the water level on side A will
    A. rise increasing the [protein].
    B. rise decreasing the [protein].
    C. lower increasing the [protein].
    D. lower decreasing the [protein].

27. Visible changes would be **LEAST** obvious when placing?
    A. Plant cells in hypotonic solutions.
    B. Plant cells in hypertonic solutions.
    C. Animal cells in hypotonic solutions.
    D. Animal cells in hypertonic solutions.

28. Thyroid glands have high concentrations of iodine due to
    A. pinocytosis.
    B. active transport.
    C. passive transport.
    D. facilitated transport.

29. The fluid mosaic model describes membranes as having
    A. protein channels within a layer of phospholipids.
    B. two layers of proteins in which phospholipids are embedded.
    C. two layers of protein separated by a layer of phospholipids.
    D. two layers of phospholipids in which specialized proteins are embedded.

30. Transport proteins in cell membranes
    A. move in and out through pores.
    B. dissolve part of the membrane to let substances through.
    C. move substances through by interacting with them.
    D. react with substances and propel the products through a membrane.

The next TWO questions refer to this sketch.

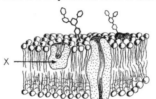

31. The sketch **BEST** represents
    A. a plant cell's cell wall.
    B. RER in an animal cell.
    C. an animal cell's nuclear envelope.
    D. an animal cell's plasma membrane.

32. Structure **X** is **MOST** likely a
    A. receptor site.
    B. feeding structure.
    C. protective feature.
    D. secretory structure.

33. Which process would be the **FIRST** to be affected by the lack of oxygen?
    A. Osmosis.
    B. Diffusion.
    C. Active transport.
    D. Facilitated transport.

WRITTEN ANSWERS:

1. Describe how these substances would **MOST** likely leave a cell.
    a. water
    b. amino acids
    c. proteins

2. Explain **THREE** different functions of membrane proteins.

3. List **THREE** factors that affect the rate of diffusion of ions through a solution.

4. A student set up the experiment illustrated below and kept it at room temperature.

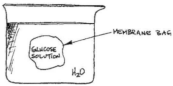

   After five minutes, the only apparent difference was that the membrane bag was swollen into a spherical shape.
    a. Explain this observation.
    b. What **TWO** statements can be made about the selective permeability of this membrane?
    c. If this same experiment were to be conducted at 5 degrees, explain how and why the results could differ?

5. a. Describe the effect of increasing the solute concentration in a cell's environment.
   b. Describe the effect of decreasing the solution concentration in a cell's environment.
   c. Explain why both of these changes could have an impact on the cell's ability to function normally.

6. Does osmosis obey the Law of Diffusion? Explain.

7. Explain how the following apparatus can be used to determine osmotic pressure.

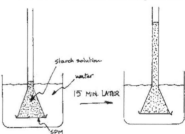

8. Four identical 0.2 g pieces a given potato were prepared, weighed and put into salt solutions. After 24 hours they were weighed again and changes in mass were noted as follows:
    • the 0.8% solution caused a loss of 0.5 g
    • the 1.5% solution caused a loss of 0.1 g
    • the 2.5% solution caused a gain of 0.2 g
    • the 3.0% solution caused a gain of 0.5 g

   Use this data to determine the concentration of salt in the potato.

UNIT E – ENZYMATICS

All organisms depend on their **metabolism** for survival. Metabolism is the combination of the biochemical reactions and pathways that occur in cells. The rate of metabolism is partially governed by the protein hormone **thyroxin**. The thyroid gland, located in the neck, accumulates iodine by active transport in order to produce thyroxin. The cells of the thyroid gland release thyroxin into the blood so it circulates throughout the body. Thyroxin attaches to receptor sites on the surfaces of body cells where it increases the rate that cells consume oxygen. This promotes ATP production by cellular respiration, thereby potentially increasing the rate of cell metabolism.

Enzymes increase the overall efficiency of metabolic reactions. The function of enzymes is dependent on the precision by which they can form temporary complexes with **substrates**. The shapes of the molecules involved allow them to combine. Anything that affects the shapes of the molecules and therefore the formation of these complexes will affect the potential for reactions.

ENZYMATIC ACTIVITY

Enzymes are proteins with tertiary (or quaternary) structure. Each one has a characteristic 3-D configuration that is a product of its unique sequence of amino acids. Molecules that react with enzymes are called

substrates. Maltose, for example, is the substrate for the enzyme, maltase. The unique structure of maltase allows it to bind with maltose and form a complex that undergoes a reaction. Maltase catalyzes the hydrolysis of maltose into its two component glucose molecules. Similarly, sucrase reacts with sucrose (substrate) to produce glucose and fructose. [Note that enzyme names generally end with "ase".]

Figure E-1. Hydrolysis of Sucrose. *This reaction shows the hydrolysis of the disaccharide sucrose by the enzyme sucrase into the products, glucose and fructose. Water is required in this reaction. In biochemical reactions it is standard to write the names of enzymes under the arrows.*

This cleaving of a single substrate into two (smaller) products is an example of **catabolism**. The opposite type of enzymatic reaction also occurs, where substrates bond together to form a single product. These reactions are **anabolic (anabolism)**. Overall, metabolism is the combination of enzymatically controlled catabolic reactions (hydrolysis) and anabolic reactions (synthesis). Many enzymatic reactions are reversible. For example, maltase not only catalyzes the hydrolysis of maltose, it also catalyzes the reverse (anabolic) reaction between glucose molecules to form maltose.

Homeostasis, the maintenance of a relatively constant set of internal body conditions, results from enzymatic reactions. Food molecules, once ingested by organisms, are digested enzymatically. As well, cells use enzymatic reactions to make the molecules they need. The conditions in which an enzymatic reaction takes place determine to what extent the reaction will proceed, or if it will take place at all. Without efficient enzyme function, an organism would not survive.

The **Lock and Key Analogy** attempts to clarify the physical relationship between an enzyme and its substrate. This analogy states that a substrate and an enzyme fit together like a key into a lock. The analogy is very useful because it stresses that each component in the reaction must have the correct shape and fit together perfectly for a reaction to occur. The portion of the enzyme that attaches to the substrate and is involved in the reaction is called the **active site**. The conformation (shape) and chemical nature of the active site results from the precise combination of the amino acids that comprises it. An active site must match the shape and chemical nature of a substrate in order to unite with it to form what is known as an **activated enzyme-substrate (E-S) complex**. It is believed that as the E-S complex forms, the substrate is forced into a slightly different shape, putting stresses on the bonds holding it together, which leads to the reaction. This aspect of the Lock and Key Analogy is known as the **induced fit hypothesis**. The reaction proceeds when the required energy exists. Throughout this process the enzymes are not significantly affected, and after the

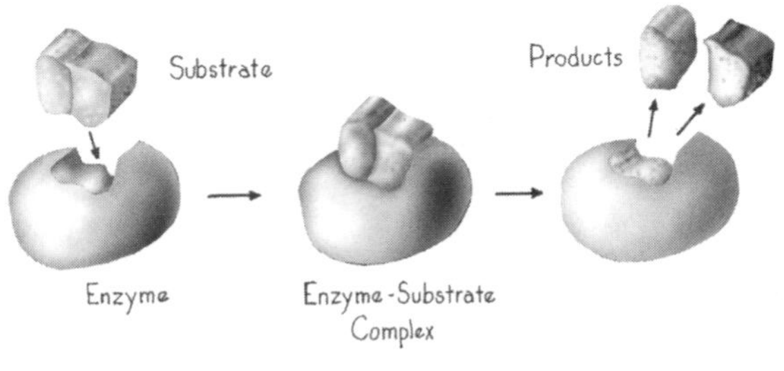

Figure E-2. Lock and Key Analogy. *Enzymes are considered to function in a lock and key manner. Each enzyme has a specific region on its surface called an active site. This region's shape matches that of its substrate. When a substrate fits into the active site, an activated enzyme-substrate complex is formed. It is at this point that the reaction (hydrolysis or synthesis) occurs and the product(s) result.*

reaction they are available to be used again. The critical part of the Lock and Key Analogy is the ability to form the E-S complex. If an activated E-S complex can't form, then no reaction can occur.

The Lock and Key Analogy can be extended. A wrong key can sometimes be fit into a lock, yet does not do anything. In "enzyme language", it is possible for a substance other than the substrate to fit into an active site. Should this happen, the configuration of this combination is not exact due to the similar, but incorrect "reactant". No reaction will proceed and the reaction that could have occurred will be blocked because a non-substrate (an **inhibitor**) is occupying the active site. In other words, a substrate-like substance **inhibits** the reaction.

E.1 CONCEPT CHECK-UP QUESTIONS:
1. How do enzymes affect metabolism?
2. What is the difference between anabolism and catabolism?
3. How important is the shape of an enzyme? Why?
4. How can the presence of inhibitors affect metabolism?

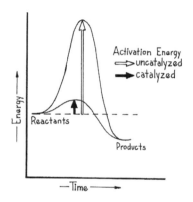

Figure E-3. Activation Energy. *Reactions require the input of energy to start. This energy is termed the activation energy. Enzymes lower the activation energy of reactions meaning catalyzed reactions can occur with less energy requirements than if they were not catalyzed.*

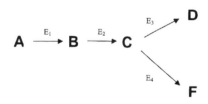

Figure E-4. Metabolic Pathway. *Many reactions in cells are actually part of longer metabolic pathways, where the product of one step is the reactant of the next step and so on. Metabolic pathways are advantageous to cells because they allow reactions to occur in smaller incremental steps involving only a portion of the whole energy change of the whole reaction. Cells can also control these pathways easier by regulating the rate at which individual steps occur.*

ATP stores energy for metabolic reactions and processes. The amount of energy required for each step is called the **activation energy**. Enzymes lower this energy requirement, thus allowing the reactions to proceed more easily and more quickly. Both the substrates and products of a reaction exist at their own energy level. The reaction is the transition between these two levels. If the overall reaction requires a net input of energy, it is termed **endothermic**. If it releases energy, it is called **exothermic** and the released energy is available to be captured in the form of ATP or given off as heat. In either case, when the reactions are enzyme-controlled, the energy changes occur in such small amounts that other cell functions are not affected. This is extremely significant for organisms. With the use of enzymes, the millions of reactions that need to occur during their life processes can occur without large energy requirements or releases.

Many enzymatic reactions exist as part of a larger **metabolic pathway**, where the product of one step becomes the substrate for a subsequent reaction and so on. Metabolic pathways have several, often many of these steps. Each step requires a different specific enzyme. The sequential nature of a metabolic pathway results in energy and electron shifts that occur in small amounts therefore having negligible impact on other aspects of metabolism. They also allow a cell to have greater regulatory control over the reactions by a mechanism called **negative feedback**. Negative feedback occurs when the concentration of the product of one step increases high enough to begin to inhibit the first step that is dedicated to its own production. In this way, the rate of these reactions is self-regulated. For example, in Figure E-4, product D could inhibit enzyme E_3, thus as the amount of D increases, the rate of production of D decreases. If this were to happen, it is likely that the simultaneous production of product F would be increasing.

To present enzymatic activity purely as a function of protein complexes is an over-simplification. Certainly, the physical structure that speeds up a reaction is a protein, but these molecules don't function in isolation. The hydrolysis of maltose, for example, involves breaking the bond holding the two monosaccharides together, the addition of water and the formation of new bonds attaching the parts of the water molecule to each of the monosaccharide products. Without maltase present, no such hydrolysis occurs. Obviously, there is an energy consideration that is not reflected in the net equation. The impact on the electron distribution in maltose as the bond is being broken and again in the glucose molecules while the new bonds are being formed also deserves consideration.

Regulating these additional features of an enzymatic reaction are the roles of **co-enzymes** and **cofactors**. Enzymatic reactions that require the addition or release of small particles like hydrogen ions use organic complexes called co-enzymes. An example of a co-enzyme is **NAD** (nicotinamide adenine dinucleotide), which is derived from niacin, one of the B vitamins. NAD (and other molecules that function similarly) transports hydrogen to or from a reaction. They are known simply as **hydrogen carriers**. The presence of cofactors like the mineral ions K^{1+} and Zn^{2+} help regulate other aspects of enzymatic reactions, such as adding stability while the electron distribution shifts.

E.2 CONCEPT CHECK-UP QUESTIONS:
1. Compare the energy consumption / production of an exothermic metabolic reaction vs. the same reaction conducted without enzymes.
2. How does the presence of enzymes affect activation energy?
3. Describe negative feedback.
4. Contrast co-factors vs. co-enzymes in terms of the type of substance they are and their role in enzymatic reactions.

FACTORS AFFECTING ENZYMES

The unique structural configuration of a protein can be changed. Recall the various factors that can alter protein structure (i.e., **denature** it). The same denaturing factors plus some other conditions affect the rate of enzymatic activity.

TEMPERATURE

The hydrogen bonds that contribute to the tertiary structure of the enzyme are sensitive to temperature. If temperatures rise too much, these bonds will break and, like any other protein, the enzyme loses its tertiary structure allowing it to take on a new set of properties. Think of cooking an egg (egg white is mostly **albumin**, a protein). Cooked egg white has a different appearance than non-cooked egg white. This sensitivity to temperature is also true for the proteins that function as enzymes. Every enzyme has a "best" or **optimum temperature**. The optimum temperature is the temperature when the shape of the active site most exactly matches the shape of the substrate. In the human body this temperature is generally about 37°C. Deviations from this temperature affect the rate of enzyme activity. If the deviation is large enough, the enzymes will be too denatured (change shape; lose 3° structure) and will cease to function.

As body temperature increases, many human enzymes continue to function at a decreasing rate up to about 50°C before they are denatured. A fever of this magnitude would be lethal. At temperatures lower than 37°C, reactions slow down because of decreased **kinetic energy** (less molecular movement). Enzymes still function but at a decreasing rate as temperatures continue to drop. This is documented in the case of people who suffer from **hypothermia**. Some victims of hypothermia have lost 15 or 20°C and have been revived. The cooling down of the molecules may not disrupt any bonds just strengthen ones that are already there.

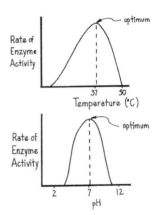

Figure E-5. Optimal Temperature and pH. *Each enzyme has an optimal temperature and pH, which allow maximum activity. For temperature, a graph of this activity is asymmetrical. At low temperatures, the reactions slow down due to decreased kinetic energy. At high temperature and pH as well as at low pH, the reactions slow down because the enzymes denature.*

pH

As with temperature, an enzyme functions best at an **optimum pH**. Deviations from this pH can result in the denaturing of the enzyme. Such is the case with the digestive system. Stomach enzymes function at a pH of about 2.5. This low pH is a result of HCl production by cells in the stomach lining. The enzymes of the small intestine function better at a pH of close to 8.4. The body has a mechanism for adjusting the pH of the food material as it moves between these organs. At the same time, if the pH of the blood were to deviate very far from its average pH of between 7.3 and 7.4, death would be imminent.

HEAVY METAL IONS

Because of their large positive nuclei, **heavy metal ions** (like Hg^{2+} and Pb^{4+}) have a great affinity for electrons. When ions like these are together with enzymes, they tend to disrupt the normal distribution of electrons in the enzymes, which affects the enzyme's shape, therefore lessening the chance that activated enzyme-substrate complexes will form and, thereby, decreasing enzymatic activity.

INHIBITORS

As was suggested in the description of the Lock and Key Analogy, the presence of **inhibitors** can affect the ability of an enzyme to bond with its substrate. **Competitive inhibitors** compete with the substrate for occupancy of the active site. The greater the **concentration** of the inhibitors relative to the concentration of the substances, the more of an affect on the rate of enzyme activity they will have. As long as an inhibitor occupies an active site, the enzyme is prevented from functioning, no activated enzyme-substrate complex can be formed, and the reaction will not occur.

There is another class of inhibitors, **non-competitive** ones, which combine with the enzyme in a location other than the active site and distort its conformation (shape). Again, this prevents the formation of the activated enzymes-substrate complex.

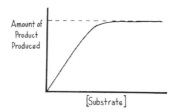

Figure E-6. Effect of Substrate Concentration.
As the substrate concentration increases in a reaction system, the rate of the reaction also increases until the concentration of the enzymes becomes the limiting factor. After this point is reached, the addition of more substrate will not increase the rate of the reaction.

SUBSTRATE AND ENZYME CONCENTRATION

Substrates are molecules that cells obtain either by transport across their membranes, or they are molecules that the cells have produced from some previous reaction. In either case, a limited amount of substrate minimizes the reaction rate. As a substrate concentration increases, the rate of the enzymatic reaction increases to the point where the cell is conducting the reaction as fast as it can because all the enzymes are "in use". Above this concentration, when substrates are in excess, the reaction will continue at a constant rate because cells do not synthesize unlimited amounts of enzymes.

Similar to the effect of substrate concentration, the concentration of the enzyme will have an impact on the reaction rate. Obviously, if there is limited enzyme available, little product can be formed.

E.3 CONCEPT CHECK-UP QUESTIONS:
1. Explain the concept of "optimum" with regards to enzyme function.
2. Why do low temperatures slow enzymatic activity?
3. What is a competitive inhibitor?
4. Compare the mechanism of action and structure of a non-competitive inhibitor with that of a heavy metal ion.

CHECK YOUR UNDERSTANDING

MULTIPLE CHOICE:

1. In a metabolic pathway
   A. products become substrates.
   B. ATP is required at each step.
   C. end products are never produced.
   D. the same enzyme is used at each step.

2. Which is **FALSE** about enzymes?
   A. Participants in reactions.
   B. Stored in a cell's active site.
   C. Very specific in their function.
   D. Denatured by high temperatures.

3. Which is **FALSE** about enzymes?
   A. Protein substances.
   B. Organic substances.
   C. Regulate metabolism.
   D. Consumed in reactions.

4. Which part of a person's diet supplies them with co-enzymes?
   A. Lipids.
   B. Proteins.
   C. Vitamins.
   D. Carbohydrates.

5. Which part of a person's diet supplies them with co-factors?
   A. Fats.
   B. Water.
   C. Minerals.
   D. Vitamins.

6. Which factor doesn't denature enzymes?
   A. pH
   B. Temperature.
   C. Concentration.
   D. Heavy metal ions.

The next question refers to this reaction.

$$\text{urea} + \text{water} \xrightarrow[]{\text{urease}} CO_2 + \text{ammonia}$$
$$\text{V} \quad \text{W} \quad\quad\quad \text{Y} \quad\quad \text{Z}$$
$$\text{X}$$

7. Which letter refers to the enzyme?
   A. V
   B. W
   C. X
   D. Y

8. Which occurs **FIRST** during hydrolysis?
   A. Water molecules used.
   B. The substrate is broken.
   C. The active site changes shape.
   D. The substrate bonds to the active site.

Use this reaction for the next two questions.

$$\text{E} + \text{S} \longrightarrow \text{"E-S"} \longrightarrow \text{E} + \text{P}$$

9. Which component is made at ribosomes?
   A. E
   B. S
   C. P
   D. None of them are.

10. Which component is temperature and pH sensitive?
    A. E
    B. S
    C. P
    D. None of them are.

11. A reaction that occurs in blood was being done in a test tube at 45 °C and at a pH of 6.5. Which change would **INCREASE** the reaction rate?
    A. Raise temperature and increase pH.
    B. Raise temperature and decrease pH.
    C. Lower temperature and increase pH.
    D. Lower temperature and decrease pH.

12. Which conditions for an enzyme reaction will produce products the slowest?

	[Substrate]	[Enzyme]
A.	Low	Low
B.	Low	High
C.	High	Low
D.	High	High

13. Energy for enzyme reactions comes from
    A. ATP.
    B. Enzymes.
    C. Substrates.
    D. Co-enzymes.

14. Which condition affecting enzymes allows them to "re-nature" once the condition is removed?
    A. A pH lower than optimum.
    B. A pH higher than optimum.
    C. A temperature lower than optimum.
    D. A temperature higher than optimum.

15. What level of structure is affected when an enzyme gets denatured?
    A. Primary.
    B. Secondary.
    C. Tertiary.
    D. Quaternary.

16. An enzyme's "ability" to recognize the appropriate substrate is based on
    A. mass.
    B. shape.
    C. reactivity.
    D. glycoproteins on their surfaces.

17. Which is **TRUE** about enzymes?
    A. They cannot be stored in cells.
    B. They never require co-enzymes.
    C. They bond onto proteins to react.
    D. High temperatures denature them.

18. Which sequence of these statements **BEST** represents an enzymatic reaction?
    1. The molecules change shape slightly.
    2. The enzyme is released unchanged.
    3. The substrates combine to form a product.
    4. The substrate fits into the active site.
    A. 1, 4, 3, 2
    B. 2, 1, 3, 4
    C. 4, 1, 3, 2
    D. 4, 2, 3, 1

19. Which is **FALSE** about enzymes?
    A. They lower activation energy.
    B. They are made out of nucleotides.
    C. They are environmentally sensitive.
    D. Their shape is specific to their function.

The next question refers to this sketch.

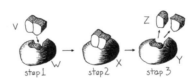

20. At which step(s) are co-enzymes and co-factors the **MOST** involved?
    A. Step 1.
    B. Step 2.
    C. Step 3.
    D. Step 1 or 2, but not Step 3.

21. The most plausible identity of **V** is
    A. ATP.
    B. Maltose.
    C. Glucose.
    D. Calcium.

The next question refers to this graph.

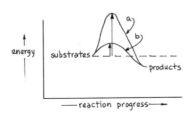

22. The difference between line "a" and line "b" represents the effect of
    A. adding enzymes.
    B. establishing optimum pH.
    C. establishing optimum temperature.
    D. increasing substrate concentration.

23. Generally, increasing enzymes in an enzymatic reaction causes the reaction to
    A. stop.
    B. reverse.
    C. speed up.
    D. slow down.

24. Which is **FALSE** about enzymes?
    A. Produced by cells.
    B. Their names end in "ase".
    C. They can be used repeatedly.
    D. React only when inside cells.

The nest questions refers to this reaction

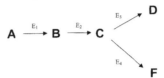

25. If substance **D** inhibits E_3, the production
    A. of substance **C** increases.
    B. of substance **D** increases.
    C. of substance **F** increases.
    D. of substances **C** and **F** increases.

26. Which is **FALSE** for enzyme reactions?
    A. Substrates get consumed.
    B. Changes occur at active sites.
    C. May involve vitamin complexes.
    D. Once started they can't be stopped.

27. If substrate concentration is increased, the amount of product formed should
    A. increase.
    B. decrease.
    C. decrease and stop.
    D. remain unchanged.

28. Which will **NOT** change the shape of an active site?
    A. Altering the pH.
    B. Adding heavy metal ions.
    C. Increasing the temperature.
    D. Adding competitive inhibitors.

29. Select the **FALSE** statement:
    A. The Kinetic Molecular Theory applies to enzymatic activity.
    B. Enzyme-catalyzed reactions are not reversible.
    C. Some enzymes work best when certain minerals are present.
    D. Enzymes can only react with a substrate with the correct shape.

30. Hg^{2+} affects enzyme activity by preventing the substrate from
    A. moving.
    B. reaching the active site.
    C. fitting into the active site.
    D. reacting once in the active site.

31. The Kinetic Molecular Theory predicts that enzyme function **DECREASES** at
    A. pH less than optimum.
    B. pH greater than optimum.
    C. temperatures less than optimum.
    D. temperatures greater than optimum.

WRITTEN ANSWERS:

1. <u>With specific reference to the Lock and Key Analogy</u>, describe the difference between substrates and competitive inhibitors.

2. During an experiment, it was discovered that increasing the amount of substrate did not speed up the reaction. Explain.

3. A test tube is prepared combining substrate solution **A** and enzyme solutions **1, 2, 3,** and **4**. The reactions that occur are:

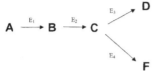

   a. Describe two ways to increase the rate of production of product **D**.
   b. A substance added to the test tube halted the production of **F**, but not **D**. Explain.

4. Two tissue samples were prepared as follows:
   Sample A: 50 g muscle tissue in a nutrient solution.
   Sample B: 50 g muscle tissue in a thyroxin and nutrient solution.

   Oxygen consumption by the samples was used to indicate of metabolic activity over a range of temperatures. The results are following:

Temperature	Consumption of oxygen (mL/hour)	
°C	Sample A	Sample B
15	6	12
25	8	16
35	12	24
45	7	14
55	2	4

   a. Sketch a graph of this data using separate lines for samples A and B.
   b. Use your graph to determine the amount of oxygen consumed per hour by Sample B at 20°C.
   c. Explain the difference observed in the results for samples A and B.
   d. Explain the variations in the results for Sample B at:
      i)   15 °C
      ii) 35 °C
      iii) 55 °C

5. Liver contains the enzyme catalase, which catalyzes the following reaction:

$$\text{hydrogen peroxide} \xrightarrow{\text{catalase}} \text{oxygen gas} + \text{water}$$
$$2H_2O_2 \qquad\qquad O_2 \qquad 2H_2O$$

   An experiment designed to measure the effect of pH on catalase activity was conducted using the following steps:
   - An equal volume of hydrogen peroxide was added to each of eight numbered test tubes at 20°C.
   - The contents of each test tube were maintained at a different pH.
   - An equal mass of liver was added to each test tube.
   - The time required to collect 10 mL of oxygen gas from each test tube was measured and recorded as shown in the table below.

Test Tube	pH of Solution	Time to Collect 10mL of O_2 (seconds)
1	5	120
2	6	90
3	7	50
4	8	30
5	9	40
6	10	60
7	11	90
8	12	140

   a. Sketch a graph that compares the time to collect 10 mL of O_2 produced to the pH of the solution.
   b. Estimate the time it would take to collect 10 mL of O_2 at a pH of 6.5
   c. Explain what causes the results observed between pH 8 and pH 12.

UNIT F - DIGESTIVE SYSTEM

The human body, like the body of any living organism, is a miraculous thing. Each of its billions of cells conducts a specific function to contribute to the whole being. These cells are arranged in tissues. There are four categories of tissues: **muscle**, which contracts and provides motion; **nerve**, which transmits impulses; **connective** (such as bone, cartilage, blood, and fat), which supports and interconnects the structures of the body and **epithelial**, which covers and protects the body and its organs. Tissues are clustered into organs, which, in turn, comprise the systems.

These tissue categories are logical at first glance - except for blood. Blood may not seem like a tissue, but it is. It is made up of cells and their extracellular fluid (plasma). It travels throughout the body and supports cells by delivering oxygen and nutrients and carrying away their wastes.

Organisms are responsive to both their internal and external environments. Through the use of different strategies, they are able to survive varying conditions. Response mechanisms like these are part of **homeostasis**, which is maintained through **negative feedback**. This is where a body responds to a change in such a way that it regains its original condition. For example, human body temperature is approximately 37°C. If it drops below this enzyme activity can be affected. The body responds by shivering (a means to generate heat) and constricting the blood flow to the surface (to avoid further temperature loss). The combined effect of these responses helps restore body temperature.

DIGESTIVE FUNCTIONS

Survival depends, in part, on an organism's ability to obtain energy for life functions. This is the role of the digestive system. The food we eat is a combination of a huge assortment of molecules. Our system exposes this food to both physical and chemical digestive processes to simplify the molecules into forms that can be absorbed into the body. Those parts of food that cannot be extracted for use have to be released. The digestive system functions are ingestion, physical and chemical digestion, absorption and defecation.

DIGESTIVE STRUCTURES

Imagine the digestive system as a long tube - some 7 meters long - contained within the body. Food enters one end, the **mouth**, and is processed throughout its length. Nutrients are absorbed and wastes are released from the other end, the **anus**. Much of the treatment that food is exposed to comes from the specializations of this tube called the **alimentary canal** (or **gastrointestinal tract**). Its muscular walls control the movement of materials through it, glands along these walls release water, enzymes, and some hormones; specialized structures absorb nutrients and so on. Two organs, the **pancreas** and the **liver**, are attached to the alimentary canal and produce secretions that are used in the digestive process. Both of these secretions enter at the **duodenum**.

The pancreas is located posterior to the stomach. It is specialized both as an **endocrine** and an **exocrine** gland. It is an endocrine (hormone-producing) gland because it produces and secretes the hormones **insulin** and **glucagon** from cells called **islets of Langerhans**. As with all hormones, these enter the blood, travel throughout the body to affect particular "target organs". Insulin lowers blood sugar levels by increasing cell's ability to use it and promoting its storage in the form of glycogen by the liver. Glucagon allows the liver to release glucose from glycogen when blood sugar levels drop. Together, these hormones help regulate the blood's level of glucose.

The pancreas is an exocrine gland because it produces the digestive juice called **pancreatic juice**, which it releases to the duodenum through the **pancreatic duct**. Pancreatic juice is a combination of six things: water, **sodium bicarbonate**, **lipase**, **trypsinogen** (which becomes **trypsin**, a **protease** enzyme), **nucleases**, and **pancreatic amylase**.

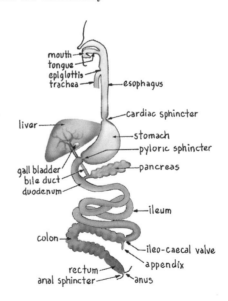

Figure F-1. Stylized Drawing of the Digestive System. *Although not anatomically accurate, this diagram shows the sequence of organs through which the components of food travel in the alimentary canal. The components of food are not technically "in" the body as long as they are in this tube. They have to be absorbed through membranes first, a process that occurs mainly in the ileum.*

The liver is the body's largest internal organ. It is located posterior to the diaphragm. The stomach is tucked up in between its lobes. The liver is richly supplied with blood, and has the following functions:

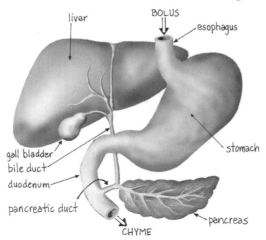

Figure F-2. Connections to the Duodenum. *The first part of the small intestines, the duodenum, receives some critical secretions that help both physical and chemical digestion. Bile, a lipid emulsifier, is produced in the liver and stored in the gall bladder. It enters through the bile duct. Pancreatic juice enters from the pancreas. This secretion contains hydrolytic enzymes, a buffer and water.*

1. ***Regulation of blood sugar levels***. Blood sugar is glucose and it should comprise about 0.1% of blood **plasma**. If the level is too high, glucose will be converted to and stored as glycogen by the liver. If it is too low, glycogen breakdown into glucose occurs (recall the roles of the pancreatic hormones, insulin and glucagon)

2. ***Production of urea***. Body cells produce ammonia (NH_3), a toxic by-product from the metabolism of amino acids and nucleotides. Ammonia is transported through blood to the liver where it is converted into the less toxic urea [$(NH_2)_2CO$]. Urea is removed from blood by the kidneys during **urine** formation.

3. ***Detoxification of blood***. The liver destroys other **toxins** in blood. A well-known example is alcohol. The liver metabolizes alcohol into fatty acids, which increases the fatty nature of liver tissue. The liver also removes and stores fat-soluble vitamins, which have toxic effects on cells when in high concentrations.

4. ***Production of bile***. Bile is an **emulsifier** of fats. As such, it mechanically breaks fat clusters into smaller pieces. This increases their surface area, thereby increasing the efficiency of the lipase enzyme, which hydrolyzes the fat into fatty acids and glycerol. Bile is stored in the gall bladder until a hormone signals its release.

5. ***Production of blood proteins (globulins)***. Examples include **albumin**, which helps maintain osmotic pressure of blood, and **fibrinogen** and **prothrombin**, which are needed for blood clotting. (These globulins are covered in greater detail in Unit G.)

6. ***Destruction of old red blood cells***. Red blood cells have an average life span of 4 months - millions die every day. Disassembled by the liver, the useful components of red blood cells are recycled. For example, iron from **hemoglobin** is returned to bone marrow for the manufacture of new red blood cells. The rest of the hemoglobin molecule is "worn out" and gets converted to **bilirubin** and **biliverdin**. These green pigment molecules are added to bile for excretion through the digestive system.

F.1 CONCEPT CHECK-UP QUESTIONS:
1. Name the four major functions of the digestive system and name an organ or structure that plays a role in each function.
2. a. What is the endocrine function of the pancreas?
   b. What is the exocrine function of the pancreas?
3. Identify two ways that the function of the pancreas is related to the function of the liver.
4. a. Categorize each of the six liver functions as: digestive, excretory, or circulatory.
   b. Why are these distinctions arbitrary?

PROCESSES OF DIGESTION

When food is ingested, it is put into the mouth and chewed. Chewing is a **physical digestion** process that increases the surface area so that **chemical digestion** can take place faster. **Salivary glands** release **saliva** through ducts (tiny tubes) into the mouth. Saliva contains water and an enzyme. The water is both a lubricant to aid in swallowing and a reactant used in the hydrolytic reactions of digestion. The enzyme, **salivary amylase (ptyalin),** breaks the bonds between the sugar molecules of **starch** to begin its chemical conversion to **maltose**.

The tongue is a muscle that facilitates chewing and rolls chewed food into "lumps" that can be swallowed. Each lump is called a **bolus**. The tongue pushes them to the back of the mouth for swallowing. Swallowing is a reflexive action resulting from contractions of the two layers of smooth muscles that line the **pharynx** and **esophagus**.

The chamber at the back of the mouth is called the pharynx. It is a common passageway for both food and air. The trachea and the esophagus start at the base of the pharynx and transport their intended contents to their respective destinations. In order to prevent the food materials from going down the trachea, its opening is covered by the **epiglottis**, a ventral flap of tissue. Muscle contractions called **peristalsis** move each bolus down the esophagus. There is a muscle constriction called the **cardiac sphincter** at the base of the esophagus that must relax and open before the food materials can enter the stomach.

The **stomach** is a large "J-shaped" organ. Its walls have three layers of muscles that **churn** the food materials over and over. The presence of food material in the stomach causes the release of **gastrin**, a hormone, which travels from the cells of the stomach walls into the blood stream. As gastrin circulates the body, it affects the stomach and causes the release of **gastric juice**. Gastric juice contains water as well as **HCl** and **pepsinogen**. Here, the boli become known as **acid chyme** (chyme, literally, means "runny"). The HCl has two functions. It creates an environment with a low pH (about 2.5) that will kill any bacterial growth that may be in the food material. It also reacts with pepsinogen and converts it into **pepsin**, a protease that converts protein into polypeptides. Protease enzymes, like pepsin are secreted in an inactive **precursor** form rather than in their final active forms because the cells that produce them would not survive their production. The inner walls of the stomach also produce mucus to protect themselves from the acid chyme. Bacterial infection of the mucosal lining by *Helicobacter pylori* can lead to an ulcer.

The **pyloric sphincter** is similar to the cardiac sphincter, except it is located at the posterior of the stomach where it controls the passage of acid chyme, a small amount at a time, into the **duodenum** (first part of the small intestine). The duodenum is specialized by the presence of **chemoreceptors**, chemical sensitive nerve endings that are able to detect the different biochemicals in the food material. In this way, the digestive system can regulate which secretions are released.

When signaled to, the gall bladder releases bile (a fat emulsifier) and the pancreas releases **pancreatic juice**. The **bicarbonate ions** (HCO_3^{1-}) from the sodium bicarbonate in pancreatic juice "over-neutralize" the acid chyme and buffer it at a pH of about 8.4 (alkaline). The enzymatic components of pancreatic juice are active at this alkaline pH: **Lipase** converts emulsified lipids into **fatty acids** and **glycerol**. **Trypsin** (a protease) breaks many peptide bonds to convert various lengths of **polypeptides** into shorter peptide units. **Pancreatic amylase** breaks remaining starch into **maltose**. Pancreatic **nucleases** convert some of the nucleic acids into nucleotides.

All of the pancreatic enzymes are active in the duodenum. The small intestine also produces and secretes its own enzymes. **Disaccharidases** like maltase and lactase complete the digestion of carbohydrates. **Peptidases** break any remaining peptide bonds in the polypeptides to produce amino acids. The small intestine also produces a variety of nucleases, which further the digestion of nucleic acids into nucleotides. Finally, nucleotidases break the nucleotides into their component molecules.

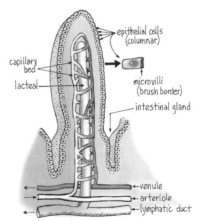

The longest part of the small intestine is called the **ileum**. The ileum is specialized to absorb the products of digestion. It has a huge surface area and is lined with finger-like projections specialized for **absorption** by active transport. These structures are called **villi**. The epithelial cells lining the villi are equipped with **mitochondria** to produce the required ATP. Internally, each villus contains a bed of loose connective tissue into which extends a capillary bed and a **lacteal**, which is an absorptive end of the lymphatic system. The products of fat digestion are reconstructed into neutral fats by the epithelial cells before they enter the lacteals, part of the lymphatic system. The rest of the products of digestion enter the blood stream.

Once the available nutrients from the food are absorbed, the remains (water plus indigestible materials like **cellulose**) pass from the ileum through the **ileo-caecal valve** (another sphincter) into the **large intestine (colon)**. The first segment of the colon is called the **caecum**. Extending down from the caecum is the **appendix**, a **vestigial structure**.

Figure F-3. Villus Structure. *These microscopic finger-like projections absorb nutrients along the inner lining of the ileum. Their epithelial cells pump the unit molecules in - amino acids, glucose, and nucleotides enter the capillaries while the products of fat digestion enter the lymphatic endings called lacteals.*

The major roles of the colon are to absorb a great amount of the water that was added to the food material all the way along the digestive system, and to house *E. coli*. *E. coli* are bacteria that metabolize some of what our bodies cannot. They live in **symbiosis** with us because they are able to obtain their nutrients from our waste materials. Their metabolism releases minerals and manufactures some vitamins and amino acids. These nutrients get absorbed along with the water into the circulatory system. The bacteria begin the decomposition of the waste materials and convert them to **feces** with a changed color, smell, and texture. The last part of the colon is the **rectum** that temporarily stores feces until **defecation**. The end of the rectum is equipped with the **anal sphincter,** which controls **defecation**.

Figure F-4. Caecum and Appendix. *A sphincter muscle called the ileo-caecal valve controls the flow of chyme from the ileum to the colon. The first lobe of the colon is the called the caecum. The appendix is a vestigial attachment to it. In herbivores, the caecum is a stomach-like pouch that permits cellulose digestion.*

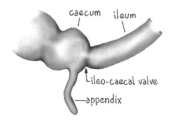

Table F-1. Summary of Digestive Enzymes

Enzyme	Source	Substrate	Product	pH
salivary amylase	salivary glands	starch	maltose	7
pepsin (a protease)	stomach	protein	polypeptides	2.5
pancreatic amylase	pancreas	starch	maltose	8.4
trypsin (a protease)	pancreas	polypeptides	peptides	8.4
lipase	pancreas	lipids	fatty acids and glycerol	8.4
nuclease	pancreas small intestine	nucleic acids	nucleotides	8.4
disaccharidases (e.g. maltase)	small intestine	disaccharides (e.g. maltose)	monosaccharides (e.g. glucose)	8.4
peptidases	small intestine	peptides	amino acids	8.4
nucleotidases	small intestine	nucleotides	sugars, bases, and phosphates	8.4

F.2 CONCEPT CHECK-UP QUESTIONS:
1. a. What is the purpose of physical digestion?
    b. Name three distinct digestive steps involving physical digestion.
2. Which two digestive enzymes function in the same way, but at different pH levels?
3. List the four enzymes from the pancreas and their substrates. (Hint: there is one for each of the four types of food chemicals.)
4. What is different about the protease enzymes pepsin and trypsin?
5. How many chemical digestion steps are required for the complete digestion of each of the four types of food biochemicals?
6. a. Name the final digestion product of each of the four types of food biochemicals.
    b. Could all of these products be correctly called monomers? Explain.
7. Describe the differential absorption of the products of digestion.
8. List the sequence of names that apply to food material as it travels through the alimentary canal.
9. a. What is the function of sphincter muscles?
    b. Name the four sphincters and identify their locations.

CHECK YOUR UNDERSTANDING

MULTIPLE CHOICE:

1. Which is **NOT** a digestive enzyme?
   A. Pepsin.
   B. Ptyalin.
   C. Gastrin.
   D. Trypsin.

2. The mitochondria in the cells lining the small intestine are required to
   A. engulf fats.
   B. digest proteins.
   C. absorb nutrients.
   D. synthesize vitamins.

3. The gall bladder's function is to
   A. store bile.
   B. synthesize lipids.
   C. produce enzymes.
   D. release pancreatic juice.

4. *E. coli* is abundant in
   A. food.
   B. feces.
   C. bile salts.
   D. stomach juices.

5. Which digestive enzyme is **NOT** correctly matched with its substrate?
   A. Lipase – fat.
   B. Pepsin – protein.
   C. Trypsin – nucleic acid.
   D. Salivary amylase – starch.

6. Glucose is stored in the liver as
   A. fat.
   B. starch.
   C. protein.
   D. glycogen.

7. Starch is first chemically digestion in the
   A. mouth.
   B. stomach.
   C. small intestine.
   D. large intestine.

8. Insulin release is *triggered* by
   A. low levels of blood sugar.
   B. high levels of blood sugar.
   C. low amounts of sugar in one's diet.
   D. high amounts of sugar in one's diet.

Refer to this sketch for the next question.

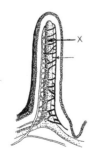

9. What enters structure **X**?
   A. Lipids.
   B. Glucose.
   C. Nucleotides.
   D. Amino acids.

10. Absorption of nutrients occurs in the
    A. liver.
    B. stomach.
    C. large intestine.
    D. small intestine.

11. The appendix attached to the
    A. stomach.
    B. gall bladder.
    C. large intestine.
    D. small intestine.

12. A function of the liver is producing
    A. urea.
    B. lipase.
    C. mucus.
    D. hydrochloric acid.

13. Removing a large section of the colon is likely to affect the
    A. release of bile.
    B. digestion of fats.
    C. absorption of glucose.
    D. water balance in the body.

14. When salivary amylase enters the stomach, it becomes
    A. basic.
    B. acidic.
    C. buffered.
    D. denatured.

15. Which region has openings for both the trachea and esophagus?
    A. throat.
    B. pharynx.
    C. stomach.
    D. epiglottis.

16. If pancreatic juice was poured over a pizza and allowed to soak for 24 hours, one would **MOST** likely find
    A. Monosaccharides and polypeptides.
    B. Amino acids, glucose, and fatty acids.
    C. Nucleotides, glucose and amino acids.
    D. Fatty acids, polypeptides, glycerol and maltose.

17. The correct match of digestive enzyme with its source is
    A. Pepsin - pancreas.
    B. Bile - gall bladder.
    C. Trypsin - stomach.
    D. Amylase - pancreas.

18. What is the **BEST** identity of **X**?

    $$\text{fat} + H_2O \xrightarrow{\text{lipase}} X + Y$$

    A. peptide.
    B. glycerol.
    C. nucleotide.
    D. amino acid.

19. Where does the chemical digestion of carbohydrates occur?
    A. mouth and large intestine
    B. mouth and small intestine
    C. mouth, stomach and pancreas
    D. mouth, stomach, small intestine

The next two questions refer to this sketch.

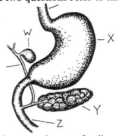

20. Which organ releases a fat-digesting enzyme?
    A. W
    B. X
    C. Y
    D. Z

21. Which organ stores food, kills bacteria, and digests protein?
    A. W
    B. X
    C. Y
    D. Z

22. Starting in the esophagus, through how many sphincters will fibre pass on its way to the small intestine?
    A. One.
    B. Two.
    C. Three.
    D. Four.

23. The cardiac sphincter prevents the movement of acid chyme from the
    A. esophagus to the mouth.
    B. stomach to the esophagus.
    C. duodenum to the stomach.
    D. colon to the small intestine.

24. The stomach is protected from HCl by
    A. Bile.
    B. Pepsin.
    C. Mucus.
    D. Bicarbonate ions.

25. Which of these can be found within the walls of the stomach, but **NOT** the chamber of the stomach?
    A. HCl.
    B. Pepsin.
    C. Gastrin.
    D. Trypsin.

26. Which structure does **NOT** produce digestive enzymes?
    A. Stomach.
    B. Pancreas.
    C. Small intestine.
    D. Large intestine.

27. The presence of food in the trachea indicates a malfunctioning
    A. pharynx.
    B. epiglottis.
    C. esophagus.
    D. cardiac sphincter.

28. Which secretion doesn't enter the duodenum?
    A. Lipase.
    B. Insulin.
    C. Trypsin.
    D. Bicarbonate ions.

29. Which enzymes are secreted in an inactive form?
    A. Lipases.
    B. Proteases.
    C. Amylases.
    D. Nucleases.

30. Pancreatic juice is
    A. alkaline (basic).
    B. a good source of vitamins.
    C. necessary for the release of insulin.
    D. unnecessary for the digestion of fat.

31. Where in the digestive system can both maltase and peptidases be found?
    A. Ileum.
    B. Stomach.
    C. Pancreas.
    D. Duodenum.

32. In which digestive organs does physical digestion occur?
    A. Mouth and stomach.
    B. Mouth and small intestine.
    C. Stomach and small intestine.
    D. Mouth, stomach and small intestine.

33. In how many of the structures listed below does peristalsis occur?
    - Esophagus
    - Stomach
    - Small intestine
    - Large intestine

    A. One
    B. Two
    C. Three
    D. Four

34. What order of terms identifies the "food" in the digestive tract?
    A. bolus, acidic chyme, feces, chyme
    B. chyme, acidic chyme, feces, bolus
    C. bolus, chyme, acidic chyme, feces
    D. bolus, acidic chyme, chyme, feces

WRITTEN ANSWERS:

1. Describe **TWO** ways the small intestine is specialized for each of these functions:
    a. transport
    b. digestion
    c. absorption

2. Describe the role of bacteria in the colon.

3. a. Name the substance that emulsifies fat in the digestive system.
    b. Why is this substance not an enzyme?
    c. Explain how emulsification assists in the chemical digestion of fat.
    d. How would a person be affected if they were unable to store this substance?

4. a. Is digestion an intracellular or an extracellular process? Explain.
    b. Where do nutrients actually enter the body? Explain.

5. Redraw and complete this table.

Substance	Source of Substance	Site of Activity	Product of Activity
Ptyalin			
Pepsin			
Glucagon			
	Pancreas		fatty acids and glycerol

6. Food material that is being digested experiences two large changes in pH as it passes through the digestive system. Identify these pH changes, what causes them and why does each one exist.

7. Detail the sequence of steps in the physical and chemical digestion and absorption of a mouthful of pure protein as it travels through the digestive system.

8. Draw a villus and explain the function of its parts.

Unit G - Circulatory System

Full appreciation of the circulatory system requires an understanding of the structural and functional relationships of its components. Physically, it is a transportation route for blood tissue through the body that is used by the respiratory system to deliver oxygen, by the digestive system to deliver nutrients and by the excretory system to prevent metabolic wastes from accumulating in tissue spaces. All cells of the body are serviced by the circulatory system in these ways during the process of capillary tissue-fluid exchange.

The source of the pressure that pushes the blood is the heart, a miraculous organ that responds automatically to the needs of the body. Blood has to pass through the heart twice for each complete circuit of the body because every time it passes through a capillary bed, the pressure is very diminished.

This unit examines the structures of the heart and blood vessels and how they function together, as well as the nature and composition of the blood and its various types of cells. The nature of the fetal circulatory system is examined along with the changes to that circulatory pattern, which have to occur at birth once the lungs become functional. The lymphatic system's close association with the circulatory system is also considered.

THE HEART

The human heart has four well-developed chambers: a pair of **atria** and a pair of **ventricles**. These chambers are arranged so that there is an atrium and a ventricle on each of the right and left side of the heart. The right side pumps blood to the lungs (**pulmonary circuit**) while the left side feeds the rest of the body systems (**systemic circuit**). This pattern of blood flow is known as a double circuit, because the blood passes through the heart twice to make one complete loop of the body.

The **atria** are the receiving chambers of the heart. The right atrium receives **deoxygenated** blood from the body through the anterior and posterior **vena cavae**. The left atrium receives **oxygenated** blood from the lungs via the **pulmonary veins**. Large flaps of muscle tissue, known as the **atrioventricular valves** (AV valves) separate the atria from the ventricles. When the heart beats, Blood flows from the atria through the AV valves and into the ventricles. Once the ventricles are full, they propel the blood towards its destination. The deoxygenated blood in the right ventricle is pumped to the lungs to get oxygenated. In contrast the oxygenated blood in the left ventricle is pumped to the rest of the body. The **septum** is a muscular wall that separates the ventricles and maintains this separation of blood.

The AV valves ensure the blood does not flow in the reverse direction back up into the atria as it otherwise could during ventricular contraction. These large valves are equipped with **chordae tendineae**, which are tiny tendons that attach the valve flaps to the interior of the ventricle walls. These tendons ensure the AV valves do not invert during the contraction of the ventricles. Another set of valves, called **semi-lunar valves** exist at the beginning of the arteries. The valve at the beginning of the **aorta** is called the **aortic valve**, where the one at the beginning of the **pulmonary trunk** is called the **pulmonary valve**. Once blood passes through these valves entering the arteries and the ventricles relax, the blood pressure is significantly lowered allowing the blood to rest against these valves, which closes them. In this way, the semi-lunar valves prevent the blood from draining back into the ventricles.

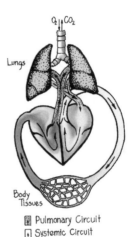

Figure G-1. Pulmonary and Systemic Circuits. *Blood makes two passes through the heart each time it travels around the body. The right side of the heart pumps it to the lungs. This is the pulmonary circuit. The activity of the left ventricle sends blood through the systemic circuit to all the other body organs.*

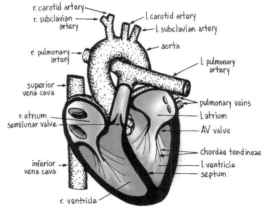

Figure G-2. Heart Structure. *The human heart has four chambers. The largest two, the ventricles are separated from each other by the septum. The right side of the heart receives deoxygenated blood from the body and pumps it to the lungs for oxygenation. The left side of the heart gets this blood and sends it to the other body organs.*

CONTROL OF HEART FUNCTION AND THE CARDIAC CYCLE

The heart contains two spots of a specialized tissue called nodes: the **SA node** (sino-atrial) and the **AV node**. Both are located in the right atrium. The SA node is along the wall of the chamber itself, where the AV node is deeper, closer to the AV valve. Nodal tissue is unique. It consists of specialized muscle cells combined with nerve cells and has the ability to contract independently of other stimuli. Nodal tissue stimulates the intrinsic nature of the heartbeat. A heart, even removed from a living organism, will continue to beat until it dehydrates, suffers a lack of ATP, or some external stimulus causes a cardiac arrest.

The SA node initiates the contraction of the heart, on the average, every 0.85 seconds (= 72 times per minute). When the atria are full the SA node stimulates their simultaneous contraction and sends a nerve impulse to the AV node causing it to respond a split second later. The contraction of the atria causes the blood pressure in them to exceed that of the ventricles, which forces the AV valves open and allows the blood to enter the ventricles. The filling time of the ventricles is coordinated with the activity of the AV node. When the ventricles are full, the AV node causes them to contract in unison to send blood out of the heart. The heightened blood pressure in the ventricles during their contraction snaps the AV valves closed. Blood is forced to enter the **aorta** through the **aortic valve** from the left side of the heart, and into the **pulmonary trunk** (through the **pulmonary valve**) on the right side of the heart. The aortic and pulmonary valves are known as **semi-lunar valves**. The **pulmonary trunk** branches to form the **pulmonary arteries** that take the blood to the lungs. The ventricles then relax and the cycle repeats itself. Because the muscle tissue of the ventricles is so massive (compared to the atria), the AV node works through the **Purkinje fibers** to coordinate the contraction of the ventricles. The Purkinje fibers are a set of nerves starting at the AV node that conduct impulses throughout the ventricles causing both ventricles to contract at once. This whole series of events, comprising one complete heart contraction, is called the **cardiac cycle**. The heartbeat has a double sound caused by the closing of the two sets of valves.

The significant role the SA node plays in initiating the heartbeat has given it the nickname **"pacemaker"**. People with irregular heartbeats may require an artificial pacemaker, which is a small device that stimulates the SA node to initiate the contraction of the heart. The SA node is connected to the brain by a nerve. When the brain perceives that the blood is getting delivered to the tissues too slowly, or if blood pressure is low, the brain will use this nerve to signal the SA node to speed up its contraction. The part of the brain that governs this activity is the **medulla oblongata**. Similarly, the medulla oblongata can slow the heart rate down. This regulation of the heartbeat is **autonomic** (i.e., not under conscious control).

BLOOD PRESSURE

Recognizing that the ventricles pump approximately 70 mL of blood each time they contract allows one to realize why the arteries must have elastic, expandable walls. The force of the blood against the blood vessel walls is known as **blood pressure**. Blood pressure is not constant. It is greater when the ventricles are contracting, actively forcing blood through the arteries. Between these contractions, the blood pressure is less - as if it were resting, waiting to be pushed farther along the artery. The term **systolic pressure** (or simply **systole**) refers to the pressure when the ventricles are contracting. **Diastolic pressure** (**diastole**) refers to the blood pressure when the ventricles are not contracting. The period of time associated with diastole includes atrial contraction and ventricular relaxation and recovery.

Blood pressure is normally measured using the **brachial artery** of the arm. A reading of 120/80 (systolic/diastolic) is normal. The units of this measurement are "mmHg".

(Normal air pressure at sea level is 760 mmHg.) High blood pressure not only puts additional pressure on the tissues that are being fed by the blood, but it is also an indication that the heart is working too hard, further straining the blood delivery system. The potential for tissue damage is greater the longer the blood pressure remains high. During physical activity, temporary elevated blood pressure is normal.

Diet and lifestyle are often to blame for a sustained elevated blood pressure. Stress can also cause high blood pressure. **Plaques** formed by fatty deposits from digested food will, if given a chance, line the inside of arteries and arterioles causing them to lose their elasticity. This is called arteriosclerosis and may lead to hardening of the arteries (atherosclerosis). This condition causes increased blood pressure (same volume of blood, but less space for it to get through). Complete blockage of the vessels will result in tissue starvation and tissue death. Blockage of the coronary arteries, which feed the heart results in death of part of the heart muscle (a coronary heart attack). High salt intake also affects blood pressure. Excessive salt will cause the body to retain water - greater fluid volume leads to greater blood pressure. **Hypertension** is the condition known as high blood pressure. Low blood pressure (**hypotension**) can also be detrimental. Proper kidney function can only be maintained if there is sufficient pressure for filtration. Luckily, the body can adjust blood pressure. Monitored by the **medulla oblongata** of the brain, the body can **dilate** (widen) arterioles thus lower blood pressure in them, or **constrict** (narrow) them to raise blood pressure.

G.2 CONCEPT CHECK-UP QUESTIONS:
1. a. In what way is the function of the SA node and the AV node the same?
   b. How are the functions of these two nodes different?
2. What events occur as part of the cardiac cycle?
3. Why is the SA node called the pacemaker?
4. a. What can cause high blood pressure? Why is high blood pressure detrimental?
   b. What can cause low blood pressure? Why is low blood pressure detrimental?

BLOOD VESSELS

There are three general kinds of blood vessels: **arteries, capillaries** and **veins**. Arteries are defined as blood vessels that carry blood *away from the heart*. In most cases they are delivering oxygenated blood to the body tissues (systemic circulation). The pulmonary artery, in contrast transports deoxygenated blood from the heart to the lungs (pulmonary circuit). Veins are defined as blood vessels that transport blood *back to the heart*. In the systemic circuit they transport deoxygenated blood; in the pulmonary circuit, they transport oxygenated blood.

Arteries and veins are structurally distinct. The walls of arteries are able to expand and receive a quantity of blood every time the heart pumps. Approximately 70 mL of blood is pumped into the aorta with each contraction of the left ventricle. The elastic nature of the thick muscular walls of the arteries allows them to rebound in size before the next heart contraction forces another volume of blood into them. The result of this action is a **pulse**, which can be felt in arteries that are close enough to the surface of the skin such as in the wrist or neck. The major arteries of the body are large and have enough tissue mass in their walls that they have to

opened   closed

be supplied with their own blood! In contrast to this, the veins are relatively thin-walled. They do not receive any sudden large volumes of blood or pressures from blood flow. Veins also contain **valves** to assist with returning blood to the heart by preventing its backward movement. There is not enough pressure in the veins to force a continuous flow of blood. Its movement is largely a result of skeletal muscle activity. The valves prevent the blood from moving backwards.

Figure G-3. Valves in Veins. *Veins are equipped with valves, flaps of tissue that open when the blood pressure on the side closest to the capillaries is great enough. Otherwise they are closed to prevent the backflow of blood.*

Another big difference between arteries and veins is their respective locations. Arteries are deeper within the tissues where they are more protected from injury and temperature loss. The pressure forcing blood through arteries is sufficiently high that if one were to be cut, the blood loss would be difficult to stop because the heart continues to pump. Venous pressure and velocity, a result of skeletal muscle activity, are very low in contrast.

The **capillaries** are the smallest of the types of blood vessel. They are so tiny that they will only allow blood cells through them one at a time. Their walls are only one cell thick (therefore lack muscle and elastic fibers). This feature facilitates the exchange of materials with the cells.

Table G-1. Contrasts between Blood Vessel Types

Basis of Contrast	Arteries	Capillaries	Veins
Structure	• thick elastic, muscular walls capable of expanding to receive blood, then regaining their normal size	• very thin walls made out of a single layer of cells	• thin walls with little muscle and elastic fibers; equipped with valves to prevent the backflow of blood
Function	• transport blood away from the heart	• connect arteries to veins; • delivery of water, O_2 and nutrients to the tissues, and • pick up water and wastes from extracellular fluids	• transport blood back to the heart
Location	• usually deep, along bones where they are protected against damage and temperature loss	• everywhere; within a few cells of each other	• often at or near the surface surrounded by skeletal muscle
Blood Pressure	• highest; • varies as the heart beats (pulse); • normal = approx. 120/80	• much lower than arteries due to branching of arteries and subdividing of blood volume into arterioles • no longer variable • high of about 35 entering a capillary bed, drops to about 15 by the venule side	• lowest because they are farthest from the heart (pump); • drops to about 0 by the time blood gets back to the heart
Blood Velocity	• highest and variable as the heart beats	• slowest as the branching of the arterioles has divided the blood into so many pathways, causing increased friction resisting its movement	• higher than capillaries; less than arteries; changes with body activity and posture; propelled by skeletal muscle activity
Total Blood Volume	• lowest	• highest; these blood vessels are the smallest, but the total volume of blood in them all over the body at any given time is the greatest. This maximizes blood surface area with body tissues for the exchange of materials	• low, but not as low as arteries; • blood spends more time in veins returning from capillary beds than it does getting to the capillary beds

Two other types of blood vessels are often described. They are **arterioles** and **venules** (small arteries and small veins, respectively). Generally, the features of arteries and veins apply to these vessels, but on a smaller scale. In addition, arterioles leading into a particular organ or region are equipped with **sphincter muscles.** When signaled, these can dilate or constrict to regulate blood pressure and flow to the intended capillary beds. Blood temperature, for example, is regulated in this manner. Increasing the blood flow to the surface tissues promotes temperature loss (by heat radiation into the air). During these times, an individual would appear "flushed". To prevent heat loss when exposed to a colder environment, the blood is prevented from going to these surface tissues and lips or fingertips may appear white as a result.

G.3 CONCEPT CHECK-UP QUESTIONS:
1. Do arteries carry only oxygenated blood and vein carry only deoxygenated blood? Explain.
2. Why is there no pulse in veins?
3. What forces blood in veins back to the heart?
4. What is the role of the sphincter muscles in the circulatory system? Where are they located?

MAJOR BLOOD VESSELS

The preceding information about blood vessels is general. When put into context, this topic begins to take shape. Some of the major blood vessels in the body and their functions are:

a. *aorta* - This is the major blood vessel carrying oxygenated blood out of the heart. It leaves the left ventricle, loops over the top of the heart creating the structure known as the **aortic arch** before descending along the inside of the backbone as the dorsal aorta. Branches from this blood vessel feed all the body systems and tissues except the lungs.

b. *coronary arteries and veins* - The very first branches off the aorta are the coronary arteries. These relatively small blood vessels can be seen on the surface of the heart. They feed the heart muscle. The heart does not receive its nutrients from the blood that travels through it. Its muscle tissue is dense and the blood is traveling though it with too much pressure and velocity. The coronary veins take the "spent blood" back to the vena cava.

c. *carotid arteries* - These branches from the aortic arch take the blood to the head (including the brain). They are highly specialized with different types of nerve endings: **chemoreceptors** that detect oxygen content, and **pressure receptors** that detect blood pressure changes. These features are essential for homeostasis. The carotid arteries run reasonably close to the surface, and the pulse in the carotids can usually be easily found along the sides of the neck.

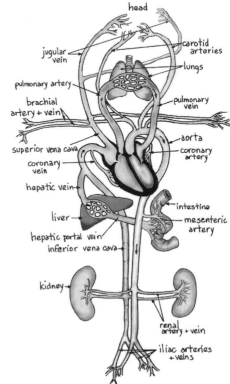

Figure G-4. Major Circulatory Pathways. *Some of the major blood vessels of the body are mapped out in this illustration. It is valuable to be able to (mentally) trace the direction of blood flow around the body.*

d. *jugular veins* - Unlike most arteries, which are paired with veins having a corresponding name, the jugular veins are the match for the carotid arteries. They conduct blood out of the head region to the anterior vena cava. The jugular veins do not contain valves. The blood flowing through them is under the influence of gravity. (When you stand on your head, your blood stays in your head!)

e. *subclavian arteries and veins* - These arteries also originate from the aorta. As their name suggests, they pass under the **clavicle** (collarbone) and branch to feed the arms (**brachial artery**) and the chest wall etc.

f. *mesenteric artery and vein* - This artery branches off the dorsal aorta. It subdivides and goes to the intestines where still smaller branches of it become the arterioles and then the capillaries that can be identified in the **villi**. While the mesenteric artery feeds the organs of the digestive system, the mesenteric vein picks up the absorbed nutrients and carries them away. This vein is reduced because of the prominence of the hepatic portal vein.

g. *hepatic portal vein* - This vein transports blood rich with nutrients directly from the intestines to the liver. Hepatic means liver; portal indicates that there is a capillary bed on both ends of it. The liver is a complex organ and has many functions. Among other things, it **detoxifies** blood, destroys aged red blood cells, and regulates the glucose concentration in the blood.

h. *hepatic vein* – When the blood leaves the liver it travels through the hepatic vein to the posterior vena cava.

i. *renal arteries and veins* - The renal arteries branch off the dorsal aorta as it passes through the lumbar (lower back) region of the body. They take blood to the kidneys while the renal veins take blood away from the kidneys and back to the posterior vena cava.

j. *iliac arteries and veins* - When the dorsal aorta gets to the pelvic region, it divides into two iliac arteries, one going down each leg. The **femoral** artery is a major branch of the iliac artery. The femoral artery services the large quadriceps muscle of the leg.

k. *anterior and posterior **vena cava*** – (also known as superior and inferior vena cava, respectively) The vena cavae collect up all the blood from the various veins of the systemic circuit and conduct it back into the right atrium. The anterior vena cava services the anterior part of the body, while the posterior services the posterior part.

l. ***pulmonary*** *arteries and veins* - Reflecting back on the difference between systemic and pulmonary circuits, all of the blood vessels listed so far are part of the systemic circuit. The pulmonary circuit is comprised of blood vessels associated with the lungs. The pulmonary trunk branches to form the right and left arteries. It conducts deoxygenated blood from the right ventricle toward the lungs while the pulmonary veins take oxygenated blood from the lungs to the left atrium.

G.4 CONCEPT CHECK-UP QUESTIONS:
1. a. Identify the arteries that branch from the aortic arch.
   b. To where does each one of these arteries conduct blood?
2. Name three differences in the composition of blood plasma found in the:
   a. hepatic portal vein and the hepatic vein
   b. renal artery and renal vein
3. Name three blood vessels that can easily be used to find a pulse. Identify the pulse location for each one.
4. What types of nerve sensitivities are located in the carotid arteries?

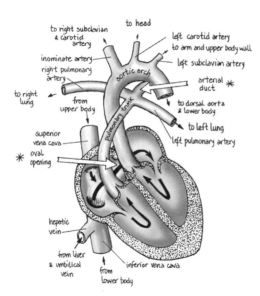

Figure G-5. Fetal Heart. *Before birth, a human heart is specialized in two ways to accommodate the non-functional pulmonary circuit. These features are the oval opening (foramen ovale), which is a valve from the right atrium into the left atrium, and the arterial duct (ductus arteriosus), which is an artery that conducts blood from the pulmonary trunk into the aorta. Both of these features allow the blood to bypass the lungs.*

FETAL CIRCULATORY SYSTEM

A fetal circulatory system is significantly different from an adult's. A **fetus** is not yet breathing and therefore does not use its pulmonary circuit. Resultantly, fetal lung tissue is too dense to allow quantities of blood to pass though it, though a small amount of blood goes to the lungs so that these tissues develop properly. Oxygenated blood is delivered to the fetus through the umbilical cord rather than the pulmonary circuit.

The **umbilical cord** has three blood vessels traveling through it. The largest one is the **umbilical vein**. It transports blood with oxygen and nutrients towards the fetal heart so it can be pumped around the developing body. Before it joins up with the posterior vena cava, it passes through the liver in the **ductus venosus (venous duct)**, which prevents the liver from processing this blood as it enters the fetus. The blood is conducted directly to the heart so it can be pumped. The blood pressure in the umbilical vein is already low and if it were to pass through liver capillaries, it might never get to the heart. Because of this, the immature tissues and organs of the fetus are exposed to any toxins that may be in the blood from the mother. The other two blood vessels in the umbilical cord are the **umbilical arteries**. These are branches off of the fetal iliac arteries. They conduct blood to the **placenta** for waste removal, gaining nutrients and oxygen.

The fetal heart has a valve called the **foramen ovale (oval opening)** between the right and left atrium. Its function is to allow the blood to bypass the lungs. It re-routes most of the oxygen-rich blood that enters the right atrium from the vena cava (via the umbilical vein) into the left atrium. A fetal heart also has a small arterial connection from pulmonary trunk to the aorta called the **ductus arteriosus (arterial duct)**. It conducts blood out of the pulmonary circuit into the systemic circuit (i.e., further allowing oxygenated blood to bypass the lungs).

Upon birth, a baby's lungs expand with the first breath. This is critical because it reduces the density of the lung tissue and, therefore, the resistance to the entry of blood into the lungs. As a result, blood enters the lungs in quantity for the first time. From there, it enters the left atrium like it never did before. This causes the closure of the foramen ovale, which grows over in a few days. The **endothelial cells** lining the ductus arteriosus are stimulated to divide due to the reduced blood flow and pressure over their surface. This vessel eventually grows closed. Its remains are called the **ligamentum arteriosus (arterial ligament)**. It serves to hold the pulmonary artery and the aorta together. The remaining unused fetal structures **atrophy** and the system becomes adult-like.

G.5 CONCEPT CHECK-UP QUESTIONS:
1. Identify the oxygenated nature of the blood and the direction of blood flow in the various blood vessels of an umbilical cord.
2. What fetal specializations allow the blood to bypass the non-functional pulmonary circuit?
3. How does a baby's first breath trigger the transformation of its circulatory system?
4. What happens to fetal circulatory tissues that are no longer needed / used?

BLOOD AND BLOOD CELLS

Blood is more than just a red liquid. It is a complex tissue that helps combat infection, regulate body temperature, and transport nutrients, O_2, and metabolic wastes. Blood is made up of three different kinds of cells transported in their extracellular fluid, the blood **plasma**. Water is the major component (over 90%) of plasma, which is the source of the blood's ability to help regulate body temperature (recall the high heat capacity of water). Plasma also transports **globulins** (proteins made by the liver), food nutrients, wastes, hormones, some dissolved gases and various ions along with the blood cells. All blood cells originate in bone marrow. They come from parent **"stem cells"** which mature in different ways to form the different types of blood cells, each with its own characteristic shape, function, and life span.

Red blood cells (erythrocytes) are continuously being produced by the bone marrow. As they develop they become biconcave in shape, which allows them to individually slip through capillaries. Their cell membrane also becomes highly specialized and equipped with unique proteins like hemoglobin and carbonic anhydrase. A third feature of their development is the loss of their nucleus. Without a nucleus, they are unable to conduct many cell functions and after a relatively short life span of about 120 days, they are worn out (chemically and physically). They become inefficient at transporting O_2 (as well as CO_2, and H^{1+} under different circumstances) and the liver removes them from the blood.

When **white blood cells (leukocytes)** mature, they specialize into several distinct types. Some develop multi-lobed nuclei and have visible granules in their cytoplasm - features lacking in others. Some specialize to conduct phagocytosis, others produce and release **antibodies**, still others release other proteins - all associated with their specific role in combating infection.

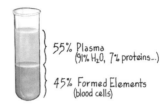

Figure G-6. Components of Blood. *When whole blood is centrifuged, its components separate according to mass. The largest portion is plasma, which is mostly water. The formed elements, or blood cells, are more massive. Most of these are the iron-containing red blood cells (most massive due to hemoglobin).*

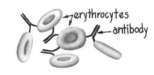

Figure G-7. Agglutination. *Agglutination results from the interaction between antibodies (products of white blood cells) and antigens, which are surface features of animal cells. When antibodies encounter matching antigens, as with foreign red blood cells (illustrated), they clump together to form a mass, which is destroyed by white blood cell activity. Large-scale agglutination can lead to death because it can plug up circulatory routes.*

Table G-2. Differences between Blood Cells

Feature	Red Blood Cell	White Blood Cell	Platelet
Structure and size	biconcave, anucleate; 7μm diameter	amoeboid; variable; 14 μm	cell pieces; tiny
Function	transport	fight infection	blood clotting
Origin	bone marrow	bone marrow; lymph tissue	bone marrow
AKA	erythrocyte	leukocyte	thrombocyte

Antibodies are Y-shaped protein complexes that are produced by white blood cells called lymphocytes. Antibodies have **receptor sites**. When they come in contact with molecules (**antigens**) that complement their structure, the two molecules bond together. One's own cells have antigens that are used for self-recognition. Generally, people do not produce antibodies that match their own antigens, but rather the antigens of foreign cells (possibly pathogenic cells). Antibodies will bind to them and cause them to clump together, or **agglutinate**. Other white blood cells will destroy the agglutinated cells by phagocytosis and/or enzymatic activity. This is part of the activity of the immune system.

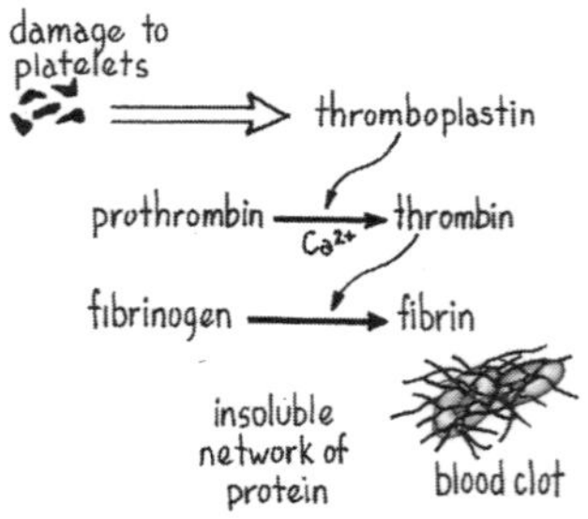

Platelets (**thrombocytes**) initiate a series of reactions that leads to the formation of a **blood clot**. Damage to platelets such as impact, or rubbing against the rough edge of a blood vessel that has been damaged, triggers them release an enzyme called **thromboplastin**. Thromboplastin catalyzes the conversion of **prothrombin** (a blood protein produced by the liver) into **thrombin**. This reaction requires the presence of Ca^{2+}. Thrombin promotes another reaction converting **fibrinogen** (another globulin) into **fibrin**. All of the proteins involved in the blood clotting reactions, with the exception of fibrin, are soluble. Fibrin is insoluble and it traps blood cells to form a blood clot.

Figure G-8. Blood Clotting Reactions. *Preventing blood loss by forming blood clots is a function of platelets. When stimulated (generally due to damage caused by cuts or tears in blood vessels), platelets release an enzyme that initiates a cascade of reactions ending with the formation of fibrin, an insoluble protein. Fibrin collects at the site of the injury trapping blood cells and sealing the wound.*

G.6 CONCEPT CHECK-UP QUESTIONS:
1. a. Name a function for each of the three types of blood cells.
   b. Name two functions of the plasma portion of blood.
2. How does the system ensure that oxygen transport is as efficient as possible?
3. a. Describe the sequence of events and function of agglutination.
   b. Describe the sequence of events and function of clotting.
4. Why are the functions of blood sensitive to changing temperatures and pH?

CAPILLARY TISSUE - FLUID EXCHANGE

Once born, a person's blood is oxygenated as it passes through their lung tissue. In this process oxygen, because its concentration is higher in inhaled air, diffuses through thin walled **alveoli**, through the capillary walls and into the blood where it bonds onto hemoglobin (Hb) on red blood cells. The oxygenated blood from the lungs enters the heart through the left atrium, gets passed to the left ventricle and pumped to body tissues.

The blood reaches a capillary bed through an arteriole and Hb releases O_2 because of different conditions (slightly warmer and less alkaline). By this time, the blood pressure has decreased to an average of about 35 mmHg (less than one third of the pressure when it left the heart). This blood pressure is sufficient enough to force water, O_2 and nutrients through the thin capillary walls out of plasma, and into the extracellular fluid in tissue spaces. Oxygen and nutrients then move into cells (by diffusion and facilitated transport, respectively) where they are used for cell respiration and other metabolic processes. The larger components of blood stay in blood (they are too big to get out). This includes all the cells and globulins (plasma proteins) like prothrombin, fibrinogen, albumin etc.

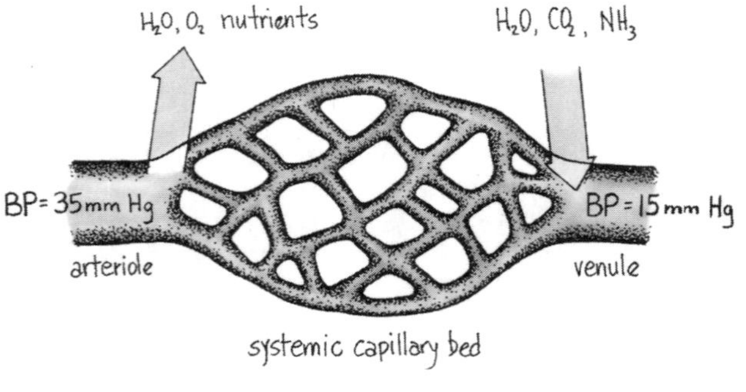

Because of these proteins, the blood in a capillary is **hypertonic** compared to the surrounding tissue, which generates an osmotic imbalance. Osmotic pressure exceeds blood pressure at the venule side of a capillary bed, which directs water back into the blood. CO_2, a by-product of cellular respiration in the cells, is more concentrated in tissue spaces than in plasma therefore diffuses back into the blood along with the water. The same is true for the metabolic waste **ammonia**, which enters plasma as well.

Figure G-9. Capillary Tissue - Fluid Exchange. *A major function of the masses of capillary beds in tissues is the delivery of oxygen and nutrients and the accumulation of wastes back into blood for eventual excretion. The oxygen and nutrient delivery occurs because of the force of the blood relative to the ability of the capillary walls to resist it. Water is pushed out carrying these things with it. This creates an osmotic gradient, which draws the water back in, this time carrying the wastes.*

Normally, this exchange results in no change in total blood volume. The exception might be if histamines were present in large quantities. Histamines are proteins released from a white blood cell called a basophil in response to damage (such as a bee sting). Histamines alter the permeability of the capillaries allowing extra fluid from plasma to enter the tissue spaces. This extra fluid does not get reabsorbed right away resulting in swelling (**edema**). Physiologically, the swelling is a defense mechanism because it dilutes whatever caused the damage.

THE LYMPHATIC SYSTEM

The **lymphatic system** has already been encountered in relation to the absorption of fatty acids and glycerol by the **lacteals** (specialized **lymph capillaries**) in the digestive system. The entire lymphatic system is a set of thin-walled vessels that start in tissue spaces throughout the body. The lymph capillaries drain excess fluids and whatever else they can from extracellular spaces and transport this fluid through a cleansing process controlled by **lymph nodes**, through **lymph ducts** back into the circulatory system. The fluid in the lymphatic system is called **lymph**.

As with veins, there is insufficient pressure in the lymphatic system to force lymph along. The one-way movement of this fluid is dependent on valves and skeletal muscle activity. This system joins the circulatory system near where the right subclavian vein joins the anterior vena cava.

G.7 Concept Check-up Questions:
1. How many cell membranes will an oxygen molecule pass through when it diffuses from air into plasma?
2. a. How is oxygen transported to systemic capillaries?
   b. Why is oxygen released from its "transport mechanism" when it gets to systemic capillaries?
3. Why do water and other substances move out of plasma and into tissue spaces at the arteriole end of a capillary bed, yet into plasma from the tissue spaces at the venule end?
4. a. How can capillary fluid exchange contribute to tissue swelling?
   b. How does this swelling get reduced?

CHECK YOUR UNDERSTANDING

MULTIPLE CHOICE:

1. The blood vessel serving the heart is the
   A. systemic artery.
   B. coronary artery.
   C. pulmonary artery.
   D. left carotid artery.

2. The co-factor required for blood clotting is
   A. K^{1+}
   B. Ca^{2+}
   C. Fe^{2+}
   D. Na^{1+}

3. Which **FIRST** receives blood from the renal vein?
   A. Aorta.
   B. Renal artery.
   C. Pulmonary artery.
   D. Hepatic portal vein.

4. Agglutination differs from clotting as it
   A. occurs in plasma.
   B. involves proteins and cells.
   C. causes cells to cluster together.
   D. is a function of the immune system.

5. Successful closure of the ductus arteriosus
   A. increases blood pressure in the aorta.
   B. decreases blood volume to the lungs.
   C. allows blood to bypass the coronary arteries.
   D. separates oxygenated from deoxygenated blood.

6. Which vessel's blood has the **LEAST** O_2?
   A. Aorta.
   B. Renal artery.
   C. Hepatic vein.
   D. Pulmonary vein.

7. Which are needed for blood clotting?
   A. Leukocytes and globulins.
   B. Leukocytes and erythrocytes.
   C. Thrombocytes and globulins.
   D. Leukocytes and thrombocytes.

8. Water moves into the venule end of capillaries as a result of
   A. Diffusion.
   B. Blood pressure.
   C. Active transport.
   D. Osmotic pressure.

9. Red blood cells are
   A. Oval.
   B. Spherical.
   C. Amoeboid.
   D. Biconcave.

10. Which is **TRUE** about systole?
    A. Atria recover and ventricles recover.
    B. Atria contract and ventricles contract.
    C. Atria recover while ventricles contract.
    D. Atria contract while ventricles recover.

11. The cardiac septum separates the
    A. left and right atria.
    B. left and right ventricles.
    C. left atrium and left ventricle.
    D. right atrium and right ventricle.

12. Contraction of the atria is controlled by
    A. SA node.
    B. AV node.
    C. Purkinje fibres.
    D. Semilunar valve.

13. Blood moves from
    A. atria to ventricles to veins.
    B. ventricles to atria to veins.
    C. atria to ventricles to arteries.
    D. ventricles to atria to arteries.

14. The chordae tendineae help prevent the backflow of blood from the
    A. atria into the ventricles.
    B. ventricles into the atria.
    C. ventricles into the aorta and pulmonary trunk.
    D. aorta and pulmonary trunk into the ventricles.

15. Which valve opens when the heart chamber producing the highest blood pressure contracts?
    A. The aortic semi-lunar valve.
    B. The left atrioventricular valve.
    C. The right atrioventricular valve.
    D. The pulmonary semi-lunar valve.

16. The heart's natural pacemaker is in the
    A. left atrium.
    B. right atrium.
    C. left ventricle.
    D. right ventricle.

17. The correct sequence of activity to coordinate a heartbeat is
    A. SA node, AV node, Purkinje fibers.
    B. AV node, SA node, Purkinje fibers.
    C. Purkinje fibers, SA node, AV node.
    D. AV node, Purkinje fibers, SA node.

18. Which arteries deliver blood to intestines?
    A. Iliac.
    B. Renal.
    C. Hepatic.
    D. Mesentaric.

19. AV valves don't invert because of the
    A. chordae tendineae.
    B. direction of blood flow.
    C. sphincter muscles in the heart.
    D. force of blood leaving the ventricles.

20. When the brain increases its stimulation of the SA node,
    A. Renal arteries and arterioles dilate.
    B. Heart rate and blood pressure decrease.
    C. Blood pressure and blood velocity increase.
    D. Production of red blood cells and platelets increase.

21. When blood moves through the AV valves, the atria
    A. and ventricles are both relaxing.
    B. and ventricles are both contracting.
    C. are contracting and the ventricles are relaxing.
    D. are relaxing and the ventricles are contracting.

22. When sphincter muscles in arterioles to the skin constrict, blood flow
    A. and blood pressure increase.
    B. and blood pressure decrease.
    C. decreases; blood pressure increases.
    D. increases; blood pressure decreases.

The next question refers to this sketch.

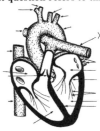

23. Which correctly identifies **X** and the composition of blood it contains?

	X	Blood Composition
A.	Vein	Oxygenated
B.	Vein	Deoxygenated
C.	Artery	Oxygenated
D.	Artery	Deoxygenated

The next question refers to this sketch.

24. The feature illustrated is characteristic of
    A. a vein.
    B. an artery.
    C. a capillary.
    D. all types of blood vessels.

25. Which describes the location and function of valves?
    A. Capillary beds - regulate the diameter of venules.
    B. Vessels with low blood pressure - prevent the backflow of blood.
    C. Vessels where blood is moving rapidly - control blood entering capillaries.
    D. Vessels carrying blood away from the heart - reduce blood pressure.

26. Which has the greatest **TOTAL** cross sectional area?
    A. Capillaries.
    B. Arterial system.
    C. Venous system.
    D. Pulmonary system.

27. Skin arterioles dilate to decrease the
    A. heart rate.
    B. blood pressure.
    C. body temperature.
    D. capillary fluid exchange.

28. The carotid arteries conduct
    A. blood to the aorta.
    B. nutrients to the heart.
    C. oxygenated blood to the head.
    D. deoxygenated blood to the lungs.

29. Glucose level should be constant in the
    A. renal artery.
    B. hepatic vein.
    C. pulmonary vein.
    D. hepatic portal vein.

30. How many heart valves would a blood cell pass through when traveling from the renal vein to the pulmonary vein?
    A. none.
    B. one.
    C. two.
    D. three.

31. Which vessels are the main pathways for blood in the systemic circuit?
    A. Aorta and vena cava.
    B. Vena cava and iliac artery.
    C. Aorta and pulmonary artery.
    D. Pulmonary artery and vena cava.

The next question refers to this sketch.

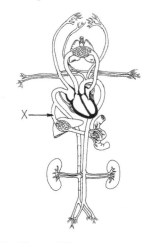

32. Identify vessel **X**.
    A. Renal vein.
    B. Hepatic vein.
    C. Inferior vena cava.
    D. Hepatic portal vein.

33. Systemic flow moves blood from the
    A. left ventricle to left atrium.
    B. right atrium to left ventricle.
    C. left ventricle to right atrium.
    D. right ventricle to right atrium.

34. Deoxygenated blood is transported by the umbilical
    A. vein and pulmonary veins.
    B. vein and pulmonary arteries.
    C. arteries and pulmonary veins.
    D. arteries and pulmonary arteries.

35. Where would a moving blood clot from the hepatic vein most likely stop?
    A. Liver.
    B. Brain.
    C. Lungs.
    D. Small intestine.

36. Blood leaving the right ventricle returns to the heart
    A. oxygenated via the left atrium.
    B. oxygenated via the right atrium.
    C. deoxygenated via the left atrium.
    D. deoxygenated via the right atrium.

37. Which is **TRUE** about the umbilical arteries?
   A. They conduct deoxygenated blood to the placenta.
   B. They receive oxygenated blood from the umbilical vein.
   C. Both of them conduct oxygenated blood into the fetus.
   D. They join together to form the umbilical vein that conducts oxygenated blood to the fetus.

38. Which of these structures allows blood to bypass the fetal lungs?
   A. Arterial duct.
   B. Venous duct.
   C. Umbilical vein.
   D. Umbilical artery.

39. Low blood pressure can result from
   A. fear.
   B. rapid pulse.
   C. dehydration.
   D. atherosclerosis.

40. The connective tissue in which cells are separated by plasma is
   A. blood.
   B. lymph.
   C. serum.
   D. extra cellular fluid.

41. Which is **NOT** transported in blood?
   A. oxygen.
   B. glucose.
   C. glycogen.
   D. fibrinogen.

42. Which are needed to fight infection?
   A. Leukocytes and proteins.
   B. Thrombocytes and proteins.
   C. Leukocytes and erythrocytes.
   D. Erythrocytes and thrombocytes.

The next question refers to this sketch.

43. The cells illustrated **CANNOT**
   A. Clot blood.
   B. Transport oxygen.
   C. Produce antibodies.
   D. Conduct phagocytosis.

44. What do white blood cells produce and release to deactivate bacteria or viruses?
   A. Platelets.
   B. Antigens.
   C. Antibodies.
   D. Hemoglobin.

45. The return of water to plasma at the venule end of a capillary bed is due to
   A. Diffusion.
   B. Blood pressure.
   C. Active transport.
   D. Osmotic pressure.

46. Excess fluid in tissue spaces will be
   A. used to form urine.
   B. removed in the form of sweat.
   C. drained away by the lymphatic system.
   D. moved into the capillary bed at a later time.

47. The lymphatic system includes
   A. lacteals and arterioles
   B. vessels with nodes and valves.
   C. the heart nodes and veins only.
   D. the heart and vessels with valves.

WRITTEN ANSWERS:

1. Describe the pathway of blood from the
   a. fingers to the toes.
   b. brain to the liver.

2. Identify and explain **ONE** relationship between each of the following.
   a. globulins and blood clotting
   b. leukocytes and agglutination

3. Describe **TWO** roles of the lymphatic system.

4. Use examples of substances to explain the role of diffusion between systemic cells and their extracellular fluids

5. Contrast diastole and systole. Describe the contraction and relaxation of the parts of the heart during each of these phases.

The next question refers to this sketch.

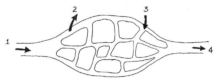

6. a. Contrast the composition of blood between positions 1 and 4.
   b. Contrast the process that occurs at position 2 with the process that occurs at 3.

7. What would be the impact on a newborn baby if the
   a. foramen ovale did not completely close?
   b. endothelial cells didn't seal the ductus arteriosus?

8. What specialization do arterioles have that help the body regulate blood flow into tissues?

UNIT H - RESPIRATORY SYSTEM

Cellular respiration, the making of ATP by mitochondria is the last part of a four-part sequence of processes related to the body's treatment and use of oxygen. The first one is **breathing**, which is inhaling (**inspiration**) and exhaling (**expiration**). Inhaling gets the oxygen into the lungs. Gases are exchanged between the blood and the lungs; oxygen goes into the blood and carbon dioxide comes out. This process is called **external respiration** as opposed to **internal respiration**, which is gas exchange between the blood and tissues in order to deliver the oxygen to cells (and get CO_2 out). Once the cells have the oxygen, their mitochondria conduct cellular respiration. CO_2 is a waste by-product that must be excreted. This unit deals thoroughly with the first three of these processes, namely getting oxygen to cells for cellular respiration.

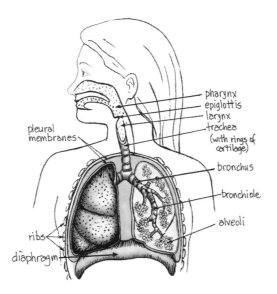

Figure H-1. Respiratory System Structures. *The structures of the respiratory system function for breathing. The contraction of the diaphragm and intercostal muscles results in a "negative pressure" in the lungs, which draws air in. External respiration occurs and exhalation follows.*

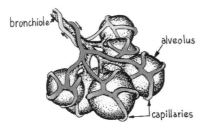

Figure H-2. Alveolar Structure. *The alveoli are the innermost reaches of the respiratory structures. It is across their surface that external respiration occurs. They are highly specialized to maximize the efficiency of this process.*

RESPIRATORY STRUCTURES

Breathing is something the body does continuously, without thinking about it. Air flows through a sequence of structures from the nose to the lungs allowing for the exchange of "oxygen-poor" air for "oxygen-rich" air. The **nostrils** are one of the major entryways into the body. They are lined with a **mucosal** cell layer, which secretes mucus. They are also equipped with nose hairs. Combined, these specializations filter and trap particulate matter. The debris that is removed from the inhaled air in this manner is discharged through the nose. A rich blood supply in the sinuses allows the presence of large numbers of white blood cells, which offer another layer of protection.

The **pharynx** is the common passageway at the back of the mouth for air and food. It ends where the **epiglottis** covers the top of the **trachea** (= glottis) leaving only the **esophagus** available for the passage of swallowed food materials. When the epiglottis is open, air is able to pass through the **larynx** (voice box) and into the trachea. The larynx contains the **vocal cords**, which are two tendons that adjust the pitch of sounds according to how taut they are. Vocalization, putting these sounds into words is a function of the mouth and tongue.

The trachea connects the pharynx to smaller air passageways leading to the lungs. It is held open and protected by the presence of C-shaped rings of **cartilage**. The open part of the "C" is at the back of the trachea, against the esophagus to facilitate swallowing. The trachea branches into **bronchi**, which also have cartilage around them to protect them against collapse. The trachea conducts air into the bronchi and the bronchi pass the air along into increasingly smaller branches of the air passageways called **bronchioles**. The continual branching of the bronchioles forms the **bronchiole tree**. The bronchioles eventually end at **alveoli**.

The alveoli are the sac-like bulges at the ends of the bronchioles. A single lung contains millions of alveoli. Combined, they provide a great deal of surface area for the diffusion of gases (external respiration). In addition to their quantity, alveoli are specialized in a number of ways:

1. The walls of the alveoli are only one cell thick which aids the diffusion of gases.
2. They have a coating of **lipoproteins** on their inner surface, which helps maintain **surface tension**. This prevents them from collapsing and sticking together during **exhalation**.
3. They are equipped with **stretch receptors**. These nerve endings are sensitive to stretch. The alveoli expand during **inhalation**. The stretch receptors send impulses to the medulla oblongata of the brain when the alveoli are full enough (stretched). This

triggers exhalation.
4. The alveolar surfaces are highly vascularized by the **pulmonary capillaries**. This ensures maximum exchange of gases.
5. They are kept moist, primarily by the circulatory system, which helps maintain their flexibility and aids gas exchange.

H.1 Concept Check-up Questions:
1. Identify and describe each of the four parts of respiration.
2. How are the nostrils specialized?
3. a. How are the large air passageways protected against collapsing?
   b. Why would it be illogical for the cartilaginous rings of the trachea to be O-shaped?
4. Explain how this sentence could help you recall the specializations of alveoli: "Thin wet capillaries stretch over numerous surfaces."

MECHANICS OF BREATHING

The respiratory centre is located in the **medulla oblongata** of the brain. It is sensitive to the concentrations of carbon dioxide and hydrogen ions in blood plasma. Both of these are toxins that result from cell metabolism and both need to be excreted. When their concentrations get to a critical level, the medulla oblongata sends nerve impulses to the **diaphragm** and the **intercostal muscles** (between the ribs) making them contract. The diaphragm moves down when it contracts. When the intercostal muscles contract, they lift the rib cage up and out. The combined effect of these motions increases the volume of the **thoracic cavity**. This creates a negative pressure in the cavity (vacuum effect) and air is drawn in through the trachea. This is inhalation (inspiration). It is an active process requiring ATP for muscle contraction.

The surfaces of the lungs are covered with a **pleural membrane**. A second outer pleural membrane coats the inside of the thoracic cavity. This double membrane not only allows the surface of the lungs to slide over the body wall easily, without abrasion, but it also seals off the thoracic cavity. A puncture to the chest wall, piercing the pleural membrane (even without damaging a lung) will result in air being drawn in through the puncture wound during inhalation, putting pressure on the surface of (instead of inside) the lung causing it to collapse. This is called **pneumothorax**.

When the stretch receptors on the surface of the alveoli detect that the alveoli are stretched open, they respond by signaling the medulla to stop the contraction of the diaphragm and intercostal muscles. When the diaphragm relaxes, it bows upwards. When the rib muscles relax, gravity pulls the rib cage down and in. These actions put pressure on the expanded thoracic cavity, which causes exhalation, the outward movement of air.

The **aortic arch** and the **carotid arteries** also contain nerve receptors called **chemoreceptors** that are sensitive to the oxygen content in blood. If it is critically low, they will help initiate the inhalation response. This is a secondary mechanism. The primary mechanism that triggers inhalation is an elevated concentration of CO_2 and H^{1+} in plasma.

CONDITIONING OF INHALED AIR

As air is drawn through the air passageways into the alveoli it is prepared in three ways. First of all, it is cleaned. The initial cleaning is by the nose hairs and mucus in the nasal passageways. The second part of the process occurs further along where debris can no longer get out through the nose. This is a role of the mucosal lining and the **cilia** along the trachea and the bronchi. Pretty well any material other than the gases of the inhaled air will get caught in the mucus. The cilia (microscopic protein filaments) are in constant motion beating the debris-laden mucus up to the pharynx to where it can be swallowed or coughed up and expectorated (spit out). The more contact air has with moist tissues that are 37°C, the closer the temperature of the inhaled air gets to 37°C. By the time the air arrives at the alveoli there is no appreciable difference between its temperature and that of the surrounding tissues. Finally, inhaled air becomes saturated with water as it passes over the mucous-lined passageways.

H.2 Concept Check-up Questions:
1. Where is the respiratory centre and what does it do?
2. a. Describe the coordinated contraction of the diaphragm and rib muscles.
   b. How does their combined contraction cause air to be drawn into the lungs?
3. What are the roles of the pleural membranes?
4. What are the three ways that inhaled air is conditioned before it gets to the alveoli?

GAS EXCHANGE

During respiration, gases are exchanged between tissues in two places. The first is between the air and blood plasma at the alveoli. This is external respiration. The other, internal respiration, occurs between blood plasma and the fluid in tissue spaces.

External respiration is the diffusion of oxygen into the pulmonary capillaries and the diffusion of carbon dioxide (and the movement of some water) into the alveoli to be exhaled. The approximate conditions in the blood at the alveoli are 37°C with a pH of about 7.38. Under these conditions **hemoglobin** combines with oxygen. Each hemoglobin molecule on a red blood cell has four bonding sites for oxygen. When the blood is leaving the alveoli, 99% of these bonding sites are occupied. The combination of hemoglobin and oxygen is called **oxyhemoglobin** (abbreviated HbO_2). It is in this manner that oxygen is transported to the tissues where internal respiration takes place.

At the tissues, the blood is slightly warmer (about 38°C) and has a slightly lowered pH of about 7.35 due to the effects of cell metabolism. Hemoglobin is very sensitive to these changes and readily releases the oxygen as the blood enters capillary beds. The oxygen diffuses into the tissue spaces along with the water that is forced from plasma (recall capillary fluid exchange). Once again, hemoglobin is available as a transport molecule.

At the venule side of the capillary bed, when water is drawn back into the blood by osmotic pressure, CO_2 and other metabolic wastes from cellular respiration also enter the blood. At these temperature and pH conditions *most* of the CO_2 reacts with H_2O in the plasma under the influence of the red blood cell enzyme **carbonic anhydrase**. This reaction temporarily forms carbonic acid, which readily dissociates.

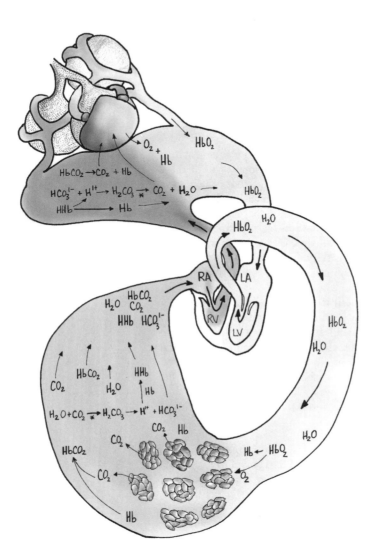

Figure H.3 External and Internal Respiration. *Chemical reactions that occur in pulmonary capillaries (external) and systemic capillaries (internal) are responsible for some of the differences in chemical composition of blood. These reactions allow CO_2 to be safely transported to the lungs for exhalation without the pH of the blood changing. They also illustrate the flexibility of hemoglobin as a transport protein.*

$$CO_2 \;+\; H_2O \;\longrightarrow\; [H_2CO_3] \;\longrightarrow\; HCO_3^{1-} \;+\; H^{1+}$$

The products, **bicarbonate ions** and hydrogen ions have different fates. Bicarbonate ions are one of the most widely known buffers in the human body. They are transported freely in blood plasma. In contrast, hydrogen ions, which would cause a decrease in the pH if transported freely in plasma, bond onto hemoglobin and are transported as **reduced hemoglobin** (HHb). In this way, blood is buffered against pH changes.

Note that *most* of the CO_2 participates in the above chemical reaction. The rest is transported in either of two ways. Some bonds onto Hb and is transported as

carbaminohemoglobin (abbreviated $HbCO_2$). Finally, a little is transported as a dissolved gas in the plasma. This completes the process known as internal respiration. The blood enters the systemic venules, returns to the right side of the heart and is pumped into the pulmonary circuit.

Blood arriving in the alveolar capillaries has the following characteristics. It is transporting bicarbonate ions and a little carbon dioxide gas. The hemoglobin is transporting either CO_2 or hydrogen. At the alveoli, as previously described, the temperature is slightly lowered and the pH is slightly higher than in the tissue spaces. These conditions result in the release of the CO_2 and hydrogen from hemoglobin. The enzyme carbonic anhydrase catalyses the reverse of the reaction that it did before:

$$HCO_3^{1-} + H^{1+} \longrightarrow [H_2CO_3] \longrightarrow CO_2 + H_2O$$

The CO_2 molecules released from this reaction, as well the CO_2 that was transported by hemoglobin and the small amount that was transported dissolved in plasma all diffuse into the alveoli and get exhaled as part of external respiration. Some of the water remains as a component of plasma. Some is exhaled as vapour. The hemoglobin is once again free to bond to oxygen and the cycle repeats itself.

H.3 Concept Check-up Questions:
1. Describe the reaction of carbonic anhydrase and explain how environmental conditions affect it.
2. Detail the transport capabilities of hemoglobin.
3. How is blood buffered against pH changes?
4. In what ways is CO_2 transported in blood. Identify the relative amount of each in deoxygenated blood.

CHECK YOUR UNDERSTANDING

MULTIPLE CHOICE:

1. HHb gets converted to HbO_2 in
   A. Liver cells.
   B. Lung capillaries.
   C. Heart ventricles.
   D. Tissue capillaries.

2. Breathing is mainly regulated by monitoring the concentration of
   A. oxygen.
   B. oxyhemoglobin.
   C. carbaminohemoglobin.
   D. carbon dioxide and hydrogen.

3. Oxygen enters blood at the alveoli by
   A. osmosis.
   B. diffusion.
   C. active transport.
   D. facilitated transport.

4. HbO_2 releases O_2 in the same region as
   A. HCO_3^{1-} is produced.
   B. NH_3 is converted to urea.
   C. CO_2 is removed from plasma.
   D. Blood pressure drops below 10 mmHg.

5. Inspiration is accomplished by
   A. relaxation of the diaphragm.
   B. relaxation of the intercostal muscles.
   C. reduced air pressure in the thoracic cavity.
   D. increased air pressure in the thoracic cavity.

6. The region of the respiratory and digestive tracts where both food and air pass is the
   A. larynx.
   B. trachea.
   C. pharynx.
   D. nasal sinus.

7. The breathing centre is located in the
   A. thalamus.
   B. cerebrum.
   C. hypothalamus.
   D. medulla oblongata.

8. Blood with the highest concentration of HHb is transported in
   A. Pulmonary veins.
   B. Coronary arteries.
   C. Hepatic portal vein.
   D. Pulmonary arteries.

9. The exchange of carbon dioxide and oxygen between the air and blood is
   A. breathing.
   B. cellular respiration.
   C. internal respiration.
   D. external respiration.

10. The correct sequence describing the outward passage of air from the body is
    A. Bronchioles, bronchi, larynx, trachea
    B. Larynx, trachea, bronchi, bronchioles
    C. Bronchioles, bronchi, trachea, larynx
    D. Trachea, larynx, bronchi, bronchioles

11. A decrease in CO_2 in blood is **MOST** closely associated with
    A. a decrease in blood pH.
    B. contraction of the diaphragm.
    C. increased bonding of O_2 to Hb.
    D. an increase in plasma temperature.

12. Blood pH remains relatively constant during internal respiration because
    A. CO_2 forms $HbCO_2$
    B. HCO_3^{1-} is produced.
    C. H_2CO_3 forms H_2O and CO_2
    D. excess H^{1+} is absorbed by leukocytes.

13. Nerve impulses from the lungs to the breathing centre are initiated by
    A. Lowering of the diaphragm.
    B. Relaxation of the diaphragm.
    C. Contraction of the rib muscles.
    D. Increased volume of the alveoli.

14. Before reaching the lungs, inhaled air is
    A. cooled and dried out.
    B. cooled and moistened.
    C. warmed and dried out.
    D. warmed and moistened.

15. As the plasma temperature decreases, the amount of CO_2 carried by hemoglobin is
    A. the same.
    B. increased.
    C. decreased.
    D. between 99 and 100%.

16. Cells use oxygen in the process called
    A. breathing.
    B. external respiration.
    C. internal respiration.
    D. cellular respiration.

17. Receptors in the respiratory centre of the brain are stimulated by
    A. low oxygen levels.
    B. high oxygen levels.
    C. low carbon dioxide levels.
    D. high carbon dioxide levels.

18. When compared to blood where Hb releases O_2, the blood where O_2 bonds to Hb is
    A. less acidic and cooler.
    B. less acidic and warmer.
    C. more acidic and cooler.
    D. more acidic and warmer.

19. HCO_3^{1-} is produced in the
    A. aorta.
    B. lung alveoli.
    C. pulmonary vein.
    D. systemic tissue capillaries.

20. A puncture in the lining of the chest cavity could result in
    A. collapse of both lungs.
    B. decreased breathing rate and fainting.
    C. collapse of the lung nearest the puncture.
    D. the leaking of gases out of the chest cavity.

21. Oxygen passes into capillaries through
    A. alveoli.
    B. bronchioles.
    C. pulmonary venules.
    D. pleural membranes.

22. Which correctly describes the muscle activity during exhalation?
    A. Diaphragm relaxes; intercostal muscles relax.
    B. Diaphragm contracts; intercostal muscles relax.
    C. Diaphragm relaxes; intercostal muscles contract.
    D. Diaphragm contracts; intercostal muscles contract.

23. Most CO_2 in blood is transported as
    A. Bicarbonate ions.
    B. Reduced hemoglobin.
    C. Carbaminohemoglobin.
    D. Dissolved gas molecules.

24. Which occurs because alveoli are moist?
    A. Bacteria are trapped.
    B. Water vapour is inhaled.
    C. Water vapour is exhaled.
    D. Gases diffuse across them easily.

25. Which occurs closest to where oxygen is released from hemoglobin?
    A. Increased pH of plasma.
    B. Increased CO_2 in plasma.
    C. Decreased body temperature.
    D. Accumulation of oxygen in tissues.

WRITTEN ANSWERS:

1. Describe **TWO** differences in conditions that hemoglobin encounters in blood and the effect of these conditions on hemoglobin's transport abilities.

2. What triggers inhalation? Exhalation?

3. Describe the interaction of the medulla oblongata, diaphragm, muscles of the rib cage, pleural membranes and lungs during inhalation and exhalation.

4. Describe the conditions, reactants, and products of the carbonic anhydrase reaction:
    a. in systemic capillaries
    b. in pulmonary capillaries

5. Describe the functions of mucus in the respiratory passageways.

6. How are alveoli specialized for their function?

UNIT I - NERVOUS SYSTEM

Survival depends, in part, on the ability to respond to changes that occur both within the body as well as in the environment around the body. Many of these responses are functions of impulses being transmitted by the nervous system. The nervous system is subdivided into the **central nervous system (CNS)** and the **peripheral nervous system (PNS)**. The CNS includes the brain and spinal cord. The PNS is comprised of two kinds of nerves. One type connects the CNS to all other parts of the body (the periphery) to collect **stimuli** for processing by the CNS. The other type conducts **responses** out of the CNS. These nerves are either part of the **autonomic nervous system (ANS)** over which we have no control (i.e., involuntary) or the **somatic nervous system (SNS)**, which we do control (i.e., voluntary).

This unit takes a cursory look at the CNS by considering some of the parts of the brain and their functions. Closer consideration is given to the spinal cord and its association with the peripheral nerves that comprise both the ANS and the SNS. As well, nerve impulses are considered both as they occur within and between neurons. The final consideration in this unit is a look at the pituitary glands and their association with the hypothalamus.

THE CENTRAL NERVOUS SYSTEM

Brain development starts in embryos as a series of enlargements at the anterior end of the neural tube, which develops into the **spinal cord**. The enlargements become the **cerebrum** and the **cerebellum** and a number of other structures that are adjacent to the cerebrum and cerebellum. There are also twelve pairs of nerves that extend out from the brain into the body regions to detect and allow responses to various stimuli. These **cranial nerves** are part of the peripheral nervous system. Cranial nerves are denoted by Roman numerals. For example, the optic nerve is cranial nerve II. The nerve going between the medulla oblongata and the heart to control heart rate is cranial nerve X, also known as the vagus nerve.

The cerebrum is the largest portion of the brain. It is divided into a right and a left **hemisphere** by the **central fissure**. At the base of this fissure is a dense tissue known as the **corpus callosum**. If one were to consider a section cut through the brain down this fissure, through the corpus callosum and on down into the spinal cord region, one would have a **sagittal** view of the brain.

The **cerebrum** is the central processing area of the brain. This is where memory is kept, where conscious thoughts are made and where a lot of non-reflexive connections and associations are made. It is a large, complex region that has its own subdivisions called lobes and areas of specialization within these subdivisions. Impulses that require processing before responding are sent to this part of the brain. Consider the example of moving to catch a ball. The stimulus (visual image of the approaching ball) causes a series of impulses that travel through clusters of sensory neurons along the optic nerve connecting the receptors of the retina to a visual association area at the back of the cerebrum. This area interprets the stimulus and sends impulses to the motor area at the front of the brain so the appropriate body movements can be made.

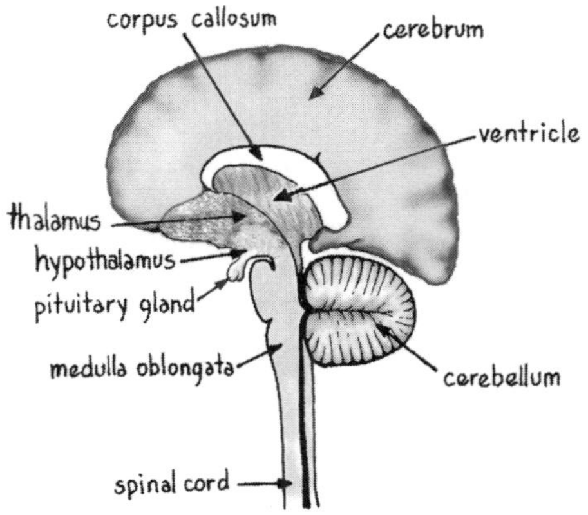

The constant streaming of sensory information from the periphery to the brain also includes information about the body's spatial orientation. These impulses are fed through the **cerebellum** and used to ensure that motor impulses (which originate near the front of the cerebrum) provide smooth movements and the body can maintain its sense of balance and coordination.

Figure I-1. Sagittal View of the Brain. *The brain develops as a set of pouches at the anterior end of the spinal cord. The two largest and most obvious regions are the cerebrum and the cerebellum. The various parts of the brain are specialized for the integration of impulses, which allows the body to coordinate thoughts and movements.*

Impulses are conducted from one side of the brain to the other through the corpus callosum. This allows the activities of one side of the brain to be coordinated with the other. Below the corpus callosum and located within the walls of the central **ventricle** (chamber) of the brain is the specialized region called the **thalamus**. Impulses traveling up the spinal cord must pass through the thalamus, which directs them to the appropriate region of the right and left cerebrum. Because of the nature of this role, the thalamus is often referred to as the **"sorting centre"**.

Another specialized region called the **hypothalamus** lies below the thalamus. The hypothalamus is known as the centre for homeostasis within the body. It constantly samples and responds to changes in the blood that travels through it in order to maintain a constant internal environment. When the hypothalamus detects that adjustments are required, it will either initiate nerve impulses resulting in responses from muscles or glands, or cause the release of hormones through its association with the endocrine system via the **pituitary gland,** which extends below it. For example, body temperature is maintained in this manner. When the hypothalamus detects that the temperature of blood is lower than it should be, it initiates the constriction of the sphincters in arterioles that conduct blood to the surface as well as the contraction of the muscles in hair follicles causing goose bumps. Both of these responses conserve body temperature and allow the core regions of the body to generate heat, thus increasing the blood temperature. The shivering response is an extension of this, which also generates heat. These are autonomic responses - they occur naturally, without thinking.

Reflexive actions are similar in that they do not require conscious thought either. Part of the brain stem called the **medulla oblongata** (at the top of the spinal cord) is dedicated to the reflexive actions that are responses to internal body stimuli. For example, it contains **receptors** that are sensitive to the conditions of the blood, such as too much CO_2 and H^{1+}

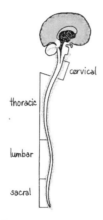

Figure I-2. Regions of the Spinal Cord. *The brain and the spinal cord comprise the CNS. While the brain functions for the integration of impulses, the spinal cord functions to route impulses either directly to effectors for reflexive actions or to the brain for integration. The spinal cord has four regions named according to the parts of the body they pass through.*

in plasma and will initiate the involuntary impulses that result in inhalation. Other reflexive actions include adjusting the heart rate, vomiting and hiccupping. A key distinction between the responses of the hypothalamus and the medulla oblongata is in terms of the speed of response. Reflexive actions like those of the medulla are relatively rapid, where the responses resulting from the activity of the hypothalamus are much slower.

The second portion of the central nervous system is the spinal cord itself. It extends from the brain down the back through the **vertebrae.** Peripheral nerves branch out into the body from between most of the vertebrae. The spinal cord has four regions. The **cervical** region corresponds to the neck. The **thoracic** region corresponds to the thoracic cavity, the **lumbar** with the lower back (abdominal) region, and the **sacral** with the tailbone area. Peripheral nerves called **spinal nerves** originate from the lower three of these regions.

The delicate CNS is surrounded by bone: the brain by the cranium and the spinal cord by vertebrae. The CNS is protected from abrasion against these bones by a set of three membranes called **meninges.** Next to the bone, the outer membrane is a tough layer called the dura mater; the one next to the nerve tissue is the pia mater. Sandwiched between these two is the vascularized arachnoid layer. Meningitis results from inflammation of the meninges.

I-1. Concept Check-up Questions:
1. List the acronyms for the different parts of the nervous system. What does each one refer to?
2. What part of the CNS do these "nicknames" refer to? Describe the role of each one.
   - a. sorting centre
   - b. homeostatic centre
   - c. reflexive centre
   - d. respiratory centre
   - e. thinking centre
   - f. coordination centre
3. What is the role of the corpus callosum?
4. What other two body systems is the hypothalamus closely associated with?

TYPES OF NEURONS

There are three kinds of **neurons,** or nerve cells: **sensory, association,** and **motor** neurons. The parts of neurons are defined by function. A **dendrite** conducts impulses towards a cell body, where an **axon** conducts impulses away from a cell body. The **cell body** contains the nucleus, which maintains the cell.

Spinal nerves are connected to the CNS either on the dorsal or the ventral side of the spinal cord creating structures called "roots". Sensory neurons have long dendrites that extend from receptors toward the CNS and pass through the **dorsal root** to the cell body, which is located in a **dorsal root ganglion**. A sensory neuron's axon is relatively short and enters the CNS from the cell body. In sharp contrast to this, an axon is the longest fibre of a motor neuron. It extends from the motor neuron's cell body, located within the CNS, through the **ventral root** and conducts impulses to an **effector** (muscle or gland) in the periphery.

Association neurons (also called **interneurons**) interconnect nerve cells. They are located entirely within the CNS. The cell bodies of motor neurons, coupled with the cell bodies of interneurons, gives the inner regions of the spinal cord a darker colour, called **gray matter**. In contrast, the outer regions of the spinal cord are called **white matter**.

Nerves, as opposed to neurons, can be seen in a dissected animal as a thin white line running along surfaces and between tissues. The dendrites of sensory neurons and the axons of motor neurons can be located in the same nerve as each other. Within one of these **mixed nerves**, there can be hundreds of the long fibres, which are covered with a fatty tissue called **myelin sheath**. Myelin is composed of **Schwann cells** that wrap around the each nerve fibre. The points between the Schwann cells are called **nodes of Ranvier**. The myelin sheath not only insulates long fibres of neurons from each other thus preventing cross-communication, but it also allows impulses to travel faster. Invertebrates, like earthworms, generally lack myelin and, therefore, are not capable of rapid nerve responses.

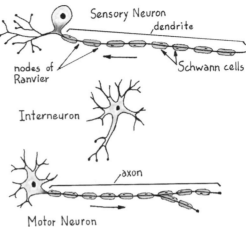

Figure I-3. Types of Neurons. *There are three different types of nerve cells (neurons). Those that conduct impulses to the CNS are called sensory neurons. Those that conduct impulses to the periphery of the body are called motor neurons. Finally, those that conduct impulses between sensory and motor nuerons are called interneurons (or association neurons). Each type is specialized in terms of structure and location.*

Table I-1. Comparisons of Types of Neurons

Bases of Comparison	Sensory Neuron	Interneuron	Motor Neuron
Structure	long dendrite, short axon	short dendrite, long or short axon	short dendrite, long axon
Function	conduct impulse to the spinal cord	interconnect sensory neurons with appropriate motor neurons	conduct impulse to an effector
Location	cell body and dendrite are outside of the spinal cord; the cell body is located along the dorsal root.	entirely within the CNS	dendrites and the cell body are located in the spinal cord; the axon is outside of the spinal cord

REFLEX ARC

The basic functional unit of the nervous system is the **reflex arc**. It usually involves all three types of neurons. The receptor could be a pain receptor in the skin. Its function is to detect stimuli that surpass the **threshold** and initiate impulses. Each impulse travels along the dendrite of the sensory neuron past the cell body, on to the axon and into the CNS to make a connection with the dendrite of an interneuron. This dendrite takes the impulse past the cell body of the interneuron to its axon. From there it makes a connection with the dendrite of the motor neuron. The impulse is

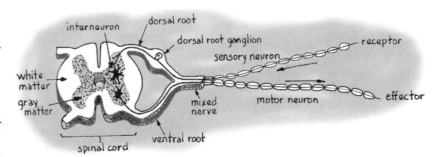

Figure I-4. Reflex Arc. *A reflexive action requires a simple neural pathway called a reflex arc. Generally the pathway includes sensory and motor neurons with association neurons to interconnect these two. Reflexes do not require integrated thought and, therefore, no interneuron is required to first conduct the impulse to the brain, though this connection will subsequently take place.*

conducted through the cell body of the motor neuron and out its axon to the effector that causes the response. The effector could be a muscle (causing movement) or a gland (resulting in a secretion). Along this pathway, there are two connections between the nerve cells within the spinal cord and another between the motor axon and the effector. These connections are called **synapses**. The brain is not involved with reflexive actions as their advantage lies in the fact that the responses occur suddenly, without any thought processes. Reflexive actions are protective mechanisms. Other interneurons will conduct impulses to the brain to alert the conscious mind that a reflexive action took place.

I.2 Concept Check-up Questions:
1. How can a sensory neuron be distinguished from a motor neuron?
2. What is the difference between a neuron, a nerve and a mixed nerve?
3. Describe the structure and function of myelin sheath.
4. a. Identify the pathway of an impulse along a reflex arc.
   b. Why are reflex arcs an advantage?

SALTATORY IMPULSE TRANSMISSION

The transmission of an impulse along a neuron occurs because of ion movement. This ion movement creates a small temporary shift in the electrical nature of the fibre. This electrical change has been studied extensively in motor axons, hence the vocabulary in explanations of impulse transmission along nerve fibres assume the fibres are axons. The process is the same in all nerve cells.

It is important to understand the electrical nature of the axon "at rest" before considering the changes that occur to it when it is conducting an impulse. There are three types of ions that are significant: Na^{1+}, which is in abundance on the outside of the axon, K^{1+} and negative ions, both of which are in abundance inside the neuron. At rest, the **axomembrane** (membrane of the axon) is not permeable to these ions. Because of the unequal distribution of the ions, the **axoplasm** (cytoplasm of the axon) has a slight negative electrical nature (-65 mV) when compared to the outside. This electrical condition is known as **resting potential**. The axomembrane contains **Na/K pumps** (proteins) that are responsible for establishing and maintaining the unequal ion distribution of resting potential.

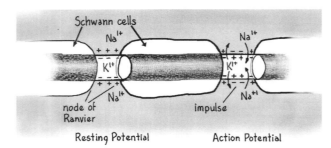

Figure I-5. Ion Movement during Impulses. *The long fibers of sensory and motor neurons are covered with myelin, a fatty sheath made out of Schwann cells. Ion movement causing an impulse takes place at the nodes of Ranvier, the spaces between the Schwann cells. Action potentials, the movement of impulses, occur with the inward movement of sodium ions followed by the outward movement of potassium ions. Active transport is required to return the ions to re-establish resting potential.*

A stimulus that surpasses the **threshold** value disrupts the axomembrane making it suddenly permeable to Na^{1+}. The sodium ions flood to the axoplasm through transport proteins, called "**sodium gates**". The inflow of Na^{1+} reverses the electrical difference across the axomembrane. The axoplasm gains a net positive charge relative to the outside. This is called **depolarization**, or **upswing** (referring to the shape the graphical representation of this process). When the electrical change to the axoplasm gets to +40 mV relative to the outside, the proteins that comprise the Na gates change their conformation once more and "close" stopping the influx of sodium ions. In a similar manner, the K gates (also transmembrane proteins) "open" allowing K^{1+} to flood to the outside restoring the normal polarity of the membrane. This is **repolarization** (or **downswing**). The Na/K pump returns the ions to their rightful places, and the neuron returns to rest.

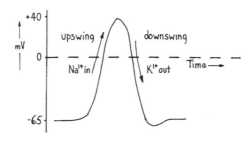

Figure I-6. Electrical Changes of an Impulse. *During resting potential, the axoplasm is negative relative to the extracellular fluid. When the sodium gates open and sodium ions flood in, the axoplasm becomes positive, which opens the potassium gates allowing potassium to flood to the outside. This restores the electrical nature, though the ions still need to be returned to their original positions.*

The magnitude of these electrical changes is always the same because the proteins in the membranes respond directly to electrical differences caused by ion movement. This means that, as long as the threshold stimulus has been reached, there will be an impulse, and each impulse is equal to each other impulse. This feature of impulses is termed **"all-or-none"** (either there will be an impulse that will be like all other impulses, or there will be none). A stronger stimulus does not mean a bigger impulse; rather it means a greater number of impulses (more neurons involved or a single neuron conducting a series of impulses).

The intense activity of ionic movement in one region of the axomembrane affects the stability of the membrane in the adjacent region. The result of this is that the permeability of these adjacent regions undergoes the same fluctuations as described above (sodium rushes in followed by potassium rushing out followed by active transport to restore the ionic positions). As this occurs in successive regions all along the fibre, the impulse moves along the fibre. In vertebrates, where most long fibres have myelin sheath around them, the Schwann cells restrict the ion movement across the axomembrane and the impulse "jumps" between successive nodes of Ranvier thus speeding up the impulse. This type of "jumping" transmission is called **saltatory transmission**.

I.3 Concept Check-up Questions:
1. Describe the electrical nature of resting potential.
2. What causes each of these components of salutatory transmission?
   a. opening the Na gates   c. opening the K gates
   b. closing the Na gates   d. closing the K gates
3. What is the role of the Na/K pump?
4. Why are all impulses the same magnitude?

SYNAPTIC TRANSMISSION

When an impulse arrives at the end of an axon, it must be transmitted to the next nerve cell (or to the muscle or gland). The axon tip or **terminus** is specialized for this type of transmission. If the receiving cell is a muscle cell, this terminus is called a **motor end plate** recognizing its flattened out shape. A terminus does not actually come in direct contact with the membrane of the receiving cell. There is a small space, termed the **synaptic gap** that must be crossed. The process by which the impulse bridges this gap is called **synaptic transmission**.

The membrane of the axon terminus is called the **pre-synaptic membrane**. The membrane on the other side of the synaptic gap is the **post-synaptic membrane**. These membranes are different in that the pre-synaptic membrane contains specialized proteins called Ca gates and encloses **synaptic vesicles** full of **neurotransmitters** manufactured by Golgi bodies in the axon. These specialized secretory vesicles are attached to the presynaptic membrane by cytoskeletal proteins. The post-synaptic membrane has protein receptor sites on it and lacks the features of the pre-synaptic membrane described above.

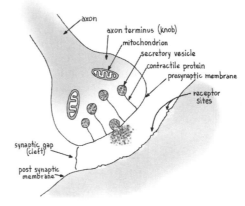

When an impulse arrives at the end of an axon, depolarization of the pre-synaptic membrane occurs (as it normally would). Depolarization causes the Ca gates to open and Ca^{2+} (present in the synaptic gap) passes to the inside of the axon. The Ca^{2+} causes the shortening of the protein filaments and results in the fusion of the vesicles of neurotransmitters with the pre-synaptic membrane. **Exocytosis** occurs and the neurotransmitters are spilled into the synaptic gap. The

Figure I-7. Synaptic Transmission. *The transmission of impulses between cells requires specialized features. On the axon side of a synaptic gap are vesicles filled with neurotransmitter molecules. The post-synaptic membrane is equipped with receptor sites into which the neurotransmitters fit in order to complete the transmission. In many instances, it seems that contractile proteins facilitate the movement and fusion of the secretory vesicles to the pre-synaptic membrane.*

neurotransmitters *diffuse* across the synaptic gap to occupy receptor sites on the post-synaptic membrane. The reception of sufficient neurotransmitters causes sodium gates to open in the post-synaptic membrane and thus the impulse is transmitted in the next cell. The synaptic gap contains **enzymes** that are specific to the neurotransmitters. The enzymes destroy the neurotransmitters, the Ca^{2+} ions are returned to the synaptic gap by **active transport**, and the synapse is returned to its original condition prior to the arrival of the next impulse. The ATP energy for the entire process comes from mitochondria in the axon.

Synaptic activity by a single axon may not necessarily result in a reaction by, the next cell. There are two kinds of neurotransmitters, **inhibitory** and **excitatory**. Each has its own effect. An inhibitory synapse makes it harder for depolarization to occur, while an excitatory neurotransmitter promotes depolarization. The summative effect of these neurotransmitters must surpass the threshold value before transmission resulting in a response can occur.

I-4 Concept Check-up Questions:
1. a. How does a presynaptic membrane differ from a postsynaptic membrane?
   b. Why can't impulses be transmitted both ways across a synapse?
2. What events would logically follow the inflowing of Na^{1+} across a postsynaptic membrane?
3. a. What is the role of enzymes present in the synaptic gap?
   b. What is the role of Ca^{2+} present in the synaptic gap?
4. Contrast the effect of an inhibitory and an excitatory neurotransmitter.

AUTONOMIC NERVOUS SYSTEM

There are two subdivisions of the autonomic nervous system. The **sympathetic** promotes active body functions while the **parasympathetic** controls the opposite, often termed **vegetative,** body functions (such as sleeping and digesting food). These subdivisions are **antagonistic** in their activity; together, they help the body maintain homeostasis. The excitatory neurotransmitter is **noradrenaline (AKA norepinephrine)**. It is destroyed by the enzyme **monoamine oxidase**. The antagonistic neurotransmitter is **acetylcholine**. It is destroyed by the enzyme **acetylcholinesterase**. For example, a sympathetic neuron releases noradrenaline to increase the heart rate while the parasympathetic neuron releases acetylcholine, which decreases the heart rate. The effect of noradrenaline (sympathetic) on the digestive system decreases its activity; the effect of acetylcholine (parasympathetic) increases its activity.

There are other distinctions between the sympathetic and parasympathetic parts of the autonomic nervous system. The motor activity of the autonomic nervous system requires two neurons to conduct an impulse to an effector where somatic neurons require only one. It follows that there must be **ganglia** along nerves outside of the central nervous system. The ganglia of the sympathetic fibres lie along side of the vertebrae where those of the parasympathetic fibres are located in the extremities, like the wrist or ankle. It follows that the parasympathetic nerves have longer **preganglionic fibres** than the sympathetic neurons. Another distinction between them is that the sympathetic fibres leave the CNS in the thoracic and lumbar regions of the spinal cord, where the parasympathetic fibres originate from the cranial and sacral regions.

Table I-2. Comparison Between Sympathetic and Parasympathetic Parts of the ANS

Bases of Comparison	Sympathetic	Parasympathetic
effect	active body function	vegetative body function
CNS origin	thoracic and lumbar	cranial and sacral
neurotransmitter	noradrenaline	acetylcholine
restoring enzyme	monoamine oxidase	acetylcholinesterase
location of motor ganglia	closer to CNS	farther from the CNS

FIGHT/FLIGHT RESPONSE

The sudden release of noradrenaline from the sympathetic neurons (as in times of fright) has a profound effect on the autonomic nervous system as described above. Stress is also interpreted by the CNS and causes the release of **adrenaline** from the **medulla** (interior) of the **adrenal glands**, located on top of the kidneys. Adrenaline (AKA epinephrine) is the hormonal version of noradrenaline; chemically they are very similar and they have related effects. Specifically, adrenaline causes the following changes:
1. Increasing the heart rate so that more blood is supplied to the body more quickly.
2. Altering blood flow patterns. Contraction of sphincter muscles in arterioles *reduces* blood flow to surface tissues (going pale) and the digestive system and allows increased blood flow to the skeletal muscles.

3. Widening the air passageways so that more air can be exchanged with each breath.
4. Contraction of skeletal and other muscles tenses the body up for action. This includes the contraction of the diaphragm. (A scared person inhales suddenly, often gasping.)
5. Contraction of the irises of the eyes to widen the pupils to maximize visual alertness.
6. Promotes the conversion of glycogen to glucose enabling increased cellular respiration generating increased ATP in cells.

NEUROENDOCRINE CONTROL

Certain responses of the body are regulated by the unified function of the hypothalamus and the pituitary gland. There are two lobes of the pituitary gland: an **anterior** and a **posterior** lobe. Both of these extend down from the hypothalamus. As blood passes through the hypothalamus, its composition and temperature stimulate various homeostatic responses. This relationship is called neuroendocrine control.

The anterior lobe of the pituitary gland releases a set of hormones that have a range of effects, such as bringing about changes associated with reproduction and skeletal growth. The posterior pituitary gland releases two hormones, ADH (which plays a role in the urinary system) and oxytocin (another hormone of the reproductive system).

The mechanism of action is different for the release of the anterior pituitary hormones than it is for the release of the posterior pituitary hormones. In the case where the hypothalamus detects that the effect of one of the hormones from the anterior pituitary is required, it releases a hormone-like substance called a "**releasing hormone**" that travels through the very short portal system that interconnects it with the anterior pituitary. A portal system is a set of blood vessels that have a capillary bed on each end (recall the hepatic portal vein). These releasing hormones trigger the secretion of the required hormones from the anterior pituitary. In contrast, the hypothalamus *produces* the hormones that are stored and released by the posterior pituitary gland. **Nerve tracts** that extend from the hypothalamus control their release. Regardless of the mechanism, all the hormones are released into the circulatory system and they travel about the body affecting specific **target organs** for which they were designed.

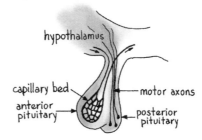

Figure I-8. Neuroendocrine Center. *The hypothalamus and the pituitary glands interact to form the neuroendocrine control center. The hypothalamus is sensitive to the body's needs and it either produces hormones, which are released from the posterior pituitary or signals the release of hormones from the anterior pituitary to accommodate these needs.*

I.5 Concept Check-up Questions:
1. Compare acetylcholine with noradrenaline.
2. a. What is the difference between noradrenaline and adrenaline?
   b. How is their activity similar?
3. How is the sympathetic response of the circulatory system different than that of the digestive system?
4. How does the mechanism of action of the anterior pituitary differ from that of the posterior pituitary?

CHECK YOUR UNDERSTANDING

MULTIPLE CHOICE:

1. Gray matter in the CNS has cell bodies of
   A. motor neurons and interneurons.
   B. sensory neurons and interneurons.
   C. motor neurons and sensory neurons.
   D. motor neurons, sensory neurons, and interneurons.

2. Rapid responses to stimuli occur due to
   A. pathways called reflex arcs.
   B. sensory neurons to the brain.
   C. transmission along the spinal cord.
   D. a large number of interneurons used.

3. From which regions of the CNS do the sympathetic neurons arise?
   A. Cranial and sacral.
   B. Lumbar and sacral.
   C. Cranial and thoracic.
   D. Thoracic and lumbar.

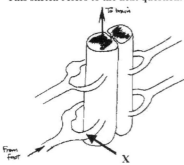

4. The sketch shows the neural pathway between the foot and the brain. If damage occurs at **X**, the person would be
   A. unable to feel pain or move the foot.
   B. able to move and feel pain in the foot.
   C. able to move the foot, but no longer feel it.
   D. able to feel foot pain, but unable to move it.

5. The part of a motor neuron in a leg is the
   A. axon.
   B. effector.
   C. dendrite.
   D. cell body.

6. Finish this analogy: nerve is to neuron as
   A. tissue is to cell.
   B. enzyme is to substrate.
   C. blood cell is to blood.
   D. protein is to peptide bond.

7. Which structures conduct impulses toward cell bodies?
   A. Sensory axon and motor axon.
   B. Sensory dendrite and motor axon.
   C. Sensory axon and motor dendrite.
   D. Sensory dendrite and motor dendrite.

8. Damage to the cerebellum could **MOST** likely cause
   A. hiccups.
   B. paralysis.
   C. lack of coordination.
   D. involuntary twitching.

This sketch refers to the next question.

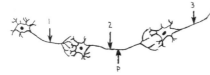

9. The axon of neuron #2 is stimulated at point **P** and a nerve impulse is generated in both directions. This impulse can only be transmitted from neuron 2 to neuron
   A. 1 by the movement of ions.
   B. 3 by the movement of ions.
   C. 1 by the release of neurotransmitters.
   D. 3 by the release of neurotransmitters.

10. Which structure is **ONLY** in the CNS?
    A. Axon of an interneuron.
    B. Axon of a motor neuron.
    C. Dendrite of a sensory neuron.
    D. Cell body of a sensory neuron.

11. Which neuron is **NOT** a participant in a spinal reflexive action?
    A. Motor neuron.
    B. Sensory neuron.
    C. Spinal interneuron.
    D. Cerebral interneuron.

This sketch refers to the next question.

12. During impulse transmission, **X** receives stimulation from
    A. a receptor.
    B. an interneuron.
    C. a motor neuron.
    D. a muscle or gland.

13. When blood gets too warm, the hypothalamus causes
    A. contraction of the spleen.
    B. involuntary muscle spasms.
    C. constriction of hair follicles.
    D. dilation of arterioles to the skin.

14. Which brain part is **BEST** associated with reflexive actions?
    A. Cerebrum.
    B. Cerebellum.
    C. Hypothalamus.
    D. Medulla oblongata.

15. Which part of the brain is mismatched with its function?
    A. Cerebrum - sensory perception.
    B. Cerebrum - integrated thought.
    C. Medulla oblongata - breathing rate.
    D. Corpus callosum – filtering impulses.

This sketch refers to the next 3 questions.

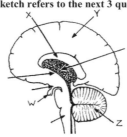

16. The structure labeled **X** is the
    A. cerebrum.
    B. cerebellum.
    C. hypothalamus.
    D. corpus callosum

17. The structure labeled **Y** is the
    A. cerebrum.
    B. cerebellum.
    C. hypothalamus.
    D. corpus callosum

18. A function of structure **Z** is to
    A. sort and relay sensory stimuli.
    B. initiate the "fight or flight" response.
    C. integrate muscle position and balance.
    D. channel information between the two hemispheres.

19. Which **MOST** correctly describes the neuroendocrine pathways?
    A. Nerves to both the anterior and posterior pituitary.
    B. Capillaries to both the anterior and posterior pituitary.
    C. Nerves to the anterior pituitary and capillaries to the posterior pituitary.
    D. Capillaries to the anterior pituitary and nerves to the posterior pituitary.

20. During resting potential, which condition exists outside the neuron?
    A. Excess K^{1+}.
    B. Excess Na^{1+}.
    C. Excess negative organic ions.
    D. Equal concentrations of Na^{1+} and K^{1+}.

21. In an axon impulses travel
    A. toward the cell body.
    B. away from the cell body.
    C. first toward, then away from the CNS.
    D. through a dendrite to a synaptic junction.

22. Which is **NOT** true of myelin?
    A. Composed of Schwann cells.
    B. Allows impulses to travel faster.
    C. Prevents cross-communication between neurons.
    D. Stores neurotransmitters for synaptic transmission.

23. Which voltage reading is mismatched?
    A. +40 mV - resting potential.
    B. 0.0 mV - sodium rushing in.
    C. -65 mV - Na/K pump functions.
    D. -10 mV - potassium rushing out.

24. Na/K pumps are responsible for establishing and maintaining
    A. action potentials.
    B. resting potentials.
    C. synaptic transmission.
    D. salt secretion from neurons.

Use this sketch for the next 3 questions.

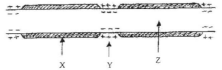

25. The shaded regions (**X**) promote the
    A. rapid response to stimuli.
    B. synthesis of neurotransmitters.
    C. integration of different stimuli.
    D. storage of ATP needed for impulse transmission.

26. Arrow **Y** is pointing at a
    A. synaptic gap.
    B. Schwann cell.
    C. myelin sheath.
    D. node of Ranvier.

27. During resting potential the potassium ions can be found in abundance at
    A. Position X
    B. Position Y
    C. Position Z
    D. Positions X and Y

28. Which event occurs first at the **LEADING** edge of an impulse?
    A. K^{1+} enters the neuron.
    B. K^{1+} leaves the neuron.
    C. Na^{1+} enters the neuron.
    D. Na^{1+} leaves the neuron.

29. Which of the following occurs during resting potential?
    A. Ca^{2+} is secreted.
    B. Na^{1+} is pumped out of axons.
    C. Axons recover neurotransmitters.
    D. Postsynaptic membranes get depolarized.

Use this graph for the next two questions.

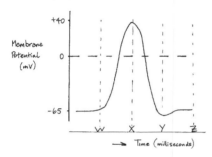

30. What occurs between time **X** and **Y**?
    A. Repolarization.
    B. Depolarization.
    C. Resting potential.
    D. Refractory period.

31. When does Na^{+1} diffuse into the neuron?
    A. Between W and X.
    B. Between X and Y.
    C. Between Y and Z.
    D. Before W and after Z.

32. The phenomenon that all impulses have the same "strength" is known as
    A. action potential.
    B. semi-conservative.
    C. refractory response.
    D. all-or-none response.

33. Na^{1+} moves to the outside of an axon by
    A. diffusion.
    B. exocytosis.
    C. active transport.
    D. facilitated transport.

34. Which is **NOT** required for impulse transmission between neurons?
    A. Enzymes.
    B. Neurotransmitters.
    C. Synaptic membranes.
    D. The presence of Ca^{2+}.

35. Mitochondria in the synaptic end of an axon provide energy for
    A. the exocytosis of neurotransmitters.
    B. pumping Na^{1+} across the synapse.
    C. pumping K^{1+} across the synapse.
    D. synthesizing receptor sites on the post-synaptic membrane.

Use this sketch for the next 2 questions.

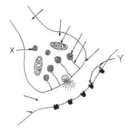

36. Structure **X** functions to transport
    A. neurotransmitters to an axon.
    B. Na^{1+} and K^{1+} to the receptors.
    C. ATP away from the mitochondria.
    D. neurotransmitters to the presynaptic membrane.

37. What do structures **Y** represent?
    A. Na/K pumps on an axon.
    B. Attachment sites for Ca^{2+} on a dendrite.
    C. Receptor sites on a postsynaptic membrane.
    D. Production sites of enzymes that destroy neurotransmitters.

38. Neurotransmitters enter synaptic gaps by
    A. diffusion.
    B. exocytosis.
    C. active transport.
    D. facilitated transport.

39. The ANS controls
    A. voluntary smooth muscles.
    B. voluntary skeletal muscles.
    C. involuntary smooth muscles.
    D. involuntary skeletal muscles.

40. Sympathetic motor axons release
    A. adrenaline.
    B. acetylcholine.
    C. noradrenaline.
    D. acetylcholinesterase.

41. Parasympathetic stimulation increases
    A. heart rate.
    B. breathing rate.
    C. smooth muscle activity.
    D. blood flow to skeletal muscles.

42. Which combination of neurotransmitter and neuron increases action in the ileum?

	neurotransmitter	neuron
A.	acetylcholine	sympathetic
B.	noradrenaline	sympathetic
C.	acetylcholine	parasympathetic
D.	noradrenaline	parasympathetic

43. Fluid taken from near a rapidly beating heart and applied to a stomach would **MOST** likely cause the stomach to
    A. speed up.
    B. slow down.
    C. secrete gastric juice.
    D. increase its oxygen uptake.

44. The ANS includes the
    A. brain and spinal cord.
    B. somatic nervous system.
    C. cerebellum and medulla oblongata.
    D. sympathetic and parasympathetic neurons.

45. Adrenaline is produced by the
    A. pancreas.
    B. adrenal cortex.
    C. adrenal medulla.
    D. medulla oblongata.

WRITTEN ANSWERS:

This sketch refers the first question.

1. Copy the above sketch and use it to
   a. **LABEL** the sensory and motor neuron.
   b. **LABEL** the axon and dendrite of the cells identified in Part a.
   c. **DRAW** arrows to indicate the direction of impulse transmission in this diagram.

2. From which specific regions of the CNS do the various autonomic nerve fibres originate?

This sketch refers to the next question.

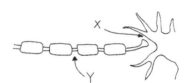

3. Nerve impulse transmission is different in area "**X**" compared to area "**Y**".
   a. Describe what occurs at **X** when an impulse is being conducted there.
   b. Describe what occurs at **Y** when an impulse is being conducted there.

4. Explain the neuroendocrine control function of the brain.

5. a. Describe the body's response to increased stimulation by the sympathetic nervous system.
   b. How does this prepare the body to respond to emergency situations?

UNIT J – URINARY SYSTEM

Excretion is an essential life function. As long as there are living tissues in a body, cells will produce metabolic wastes such as carbon dioxide and ammonia (NH_3). CO_2 is produced by cellular respiration. Body cells also deaminate amino acids (= remove the amino group from them). The 2-carbon molecule produced is metabolically significant and used by cells. The amino group is converted to ammonia, a waste by-product that is **toxic** (poisonous) when in high concentration. Ammonia is transported to the liver by the circulatory system where it is converted to **urea** [$(NH_2)_2CO$], which is less toxic. Uric acid, produced by the metabolism of nucleic acids, is another toxic nitrogenous waste. All of these wastes must be excreted because an accumulation of them becomes life threatening. CO_2 and some water are also excreted with each exhaled breath. Water is excreted through skin pores as sweat (though it is arguable that this is a secretion because of the useful effect of cooling the body off). The kidneys are the main organs of excretion. They extract urea, uric acid, excess H^{1+}, **penicillin**, **histamines**, water-soluble vitamins etc. from the blood and prepare them for excretion by urination.

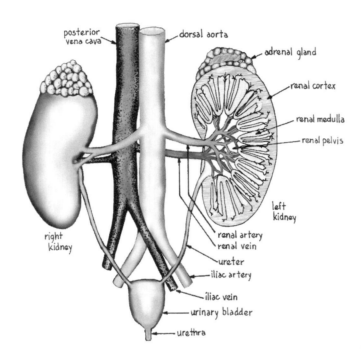

Figure J-1. Urinary System. *The urinary system consists of the kidneys, ureters, urinary bladder, and the urethra. Blood carrying urea enters the kidneys through the renal arteries, the urea is removed and concentrated in preparation for excretion, and the blood returns to the rest of the circulatory system through the renal veins.*

URINARY SYSTEM STRUCTURES

The kidneys lie in the lower, dorsal part of the abdominal cavity. They receive blood from the **renal arteries**, remove the wastes from it, and return the "clean blood" to the circulatory system. The actual waste removal from blood and production of **urine** are functions of microscopic structures called **nephrons**. Each kidney contains millions of nephrons. Urine collects in the **pelvic region** of each kidney before **peristalsis** conducts it through the **ureters** to the **urinary bladder**. When the urinary bladder is filled, the urination response is triggered, and the fluid is excreted through the **urethra**.

Nephrons are the functional units of the urinary system. These specialized tubules originate in the cortex of the kidneys. The structure of every nephron follows the same pattern. Each one begins with a "blind end" called **Bowman's capsule**. From there they twist back and forth before they loop down into the renal medulla. Each tubule loops back up to form a second twisted region before joining a common **collecting duct**. The three regions of each nephron between Bowman's capsule and the collecting duct are (in sequence) **proximal convoluted tubule, loop of Henle**, and **distal convoluted tubule**.

Blood, carrying wastes, is conducted into the cortical region of a kidney by branches of the renal artery called **afferent arterioles**. As with all blood supplies to tissues, the arterioles branch to produce capillary beds. The capillary beds associated with nephrons are highly specialized into two parts. The first part is called a glomerulus. The **glomerulus** is a tuft of delicate capillaries that are enveloped by Bowman's capsule. From the glomerulus, an **efferent** (outgoing) **arteriole** conducts blood to the second part of the capillary bed composed of **peritubular capillaries**, which extend around the rest of the nephron. Eventually, the capillaries join to form a venule, which conducts the blood back into the **renal vein**.

J.1 Concept Check-up Questions:
1. What are the excretory functions of the liver?
2. a. Name the sequence of four structures through which urine travels before it leaves the body.
   b. Identify one function of the structures names in part a.
3. What advantages does the complex shape of nephrons provide?
4. Name the sequence of blood vessels that connect the renal artery to the renal vein.

URINE FORMATION

Urine is produced by a set of processes involving the relationship between the tubular structure of the nephrons, their surrounding tissues and the blood capillaries that are associated with them. These processes are a greatly modified version of **capillary tissue-fluid exchange**. Blood entering the cortex of the kidney through the afferent arteriole is slowed down as it courses into the tiny delicate capillaries of a glomerulus. The structure of a glomerulus allows blood pressure to force small components of plasma out of the blood and into the extracellular fluid in the spaces surrounding the glomerulus. This space is occupied by a Bowman's capsule, which actually "cups around" each glomerulus. Substances thus removed from blood are forced into these porous beginnings of the nephron tubules by the continuous blood pressure. This process is the first step in urine formation. It is called **pressure filtration**. The fluid that enters the nephron tubules is called the **filtrate**. Blood cells and large molecules in plasma like **globulins** do not normally get forced into the filtrate because of their size. They stay in the remaining plasma, which leaves the glomeruli through the efferent arterioles. This blood, still oxygenated, is hypertonic to that of the afferent arteriole because of the loss of plasma and small substances that entered Bowman's capsule. Because of this differential movement of substances, **osmotic gradients** are established between the fluids (blood, filtrate, and extracellular fluid) that are separated from each other by tissues that are one-cell thick. One such gradient is between the efferent blood and the extracellular fluid. Another one is between the extracellular fluid and the filtrate that is formed by pressure filtration.

The filtrate contains some materials that are still useful and required by the body. Their small size relative to the porosity of the membranes coupled with the pressure against them allowed them to be forced into the nephron. The cells that line the proximal convoluted tubule are specially designed with **carrier proteins** to pump these needed materials back into the already hypertonic blood of the peritubular capillaries. Thus molecules like glucose, amino acids, other nutrients, etc. are returned to the blood. They are not wastes. This activity, called **selective reabsorption**, requires ATP. The removal of substances from the filtrate by selective reabsorption reduces the concentration of the filtrate. Some Na^{1+} and Cl^{1-} and water get reabsorbed at this stage. The Na^{1+} is pumped out of the filtrate into the fluid around the tubule, while the Cl^{1-} and water it follow passively. The component nature of the blood in the capillaries draws these substances back into the plasma. This accounts for a small amount of water recovery. A much greater amount occurs over the remaining nephron structure.

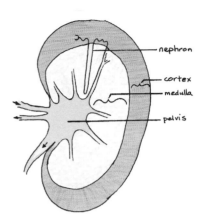

Figure J-2. Kidney Regions. *There are three regions in a kidney. The cortex is the location of the glomeruli, Bowman's capsules, and the convoluted tubules. The medulla is the region where the loop of Henle is located. The pelvis is the common area into which all the collecting ducts conduct their urine.*

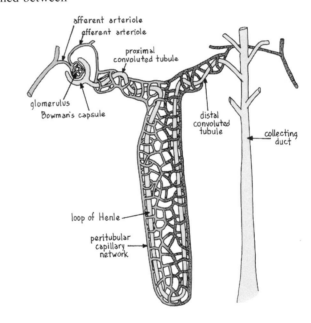

Figure J-3. Nephron Structure. *Nephrons, the structures that actually make concentrated urine have several major regions, each one specialized to conduct different aspects of the process of urine formation. The walls of the proximal and distal convoluted tubules as well as the distal portion of the ascending portion of the loop of Henle are thicker as they are equipped with numerous types of transport proteins.*

The nephron with its remaining filtrate makes three passes between the cortex and the medulla before the urine finally enters the renal pelvis. During the first pass, the filtrate goes from the proximal convoluted tubule to the medulla through the descending loop of Henle. The cells of this portion of the loop are permeable to H_2O, but not Na^{1+}. Because the loop encounters a surrounding environment with a high solute concentration as it descends into the medullary region, increasing amounts of H_2O are removed from the filtrate. This H_2O moves back into the blood due to the osmotic gradient. By the time the filtrate reaches the bottom of the loop of Henle, it has become very "salty" due to the removal of H_2O and the retention of ions.

The permeability of the ascending portion of the loop is different from the descending side. The ascending side is permeable to Na^{1+}, but not to H_2O. Because of the high Na^{1+} concentration in the filtrate relative to the fluid surrounding the tubule, Na^{1+} diffuses out of the filtrate. This feature contributes to the high solute concentration that moves water out of the descending side of the loop of Henle. As the filtrate continues to move towards the cortex, the potential for diffusion lessens as the sodium ion concentration gradient lessens. ATP is used to remove additional Na^{1+} from the filtrate. By the time the filtrate gets to the distal convoluted tubule in the cortical region of the kidney, it has been diluted by the removal of lots of Na^{1+} making it iso-osmotic to the extracellular fluids. At this point in the process, both the filtrate volume and concentration have been decreased tremendously by the removal of potential nutrients, H_2O and ions. Urea has remained throughout.

Tubular secretion follows as the filtrate passes through the distal convoluted tubule. During this process, substances that are in excess in blood (including penicillin, histamines, water-soluble vitamins etc.) can be pumped out of the peritubular capillaries around tubules into the filtrate. It is in this manner that final adjustments to the composition of the blood are made before it leaves the kidney. The pH of the blood in the distal tubule is also adjusted by the differential excretion and reabsorption of H^{1+} and HCO_3^{1-}.

In the final phase of urine formation, the filtrate is once again subjected to the increasing concentration gradient of surrounding fluids as the collecting duct passes through the medulla. Because the filtrate was diluted by the removal of Na^{1+}, the extracellular fluid in the medulla is once again hypertonic to the filtrate and additional water is withdrawn as the filtrate moves towards the renal pelvis. Throughout this process, all along the tubule, urea is minimally reabsorbed into blood because of its low **threshold level** in plasma. Some urea diffuses out of the collecting duct in the medullary region because it has become so highly concentrated. The urea that passes into the extracellular fluid contributes to the high solute concentration of the fluid (surrounding the loop) and therefore to increasing its own concentration in urine by promoting the recovery of water out of urine. These urea molecules leak back into the filtrate along the ascending side of the loop of Henle where their concentration is still relatively low. The product remaining at the end of the nephron tubules is urine, which is still mostly water even though the bulk of the water (about 99%) that entered the nephron has been reabsorbed.

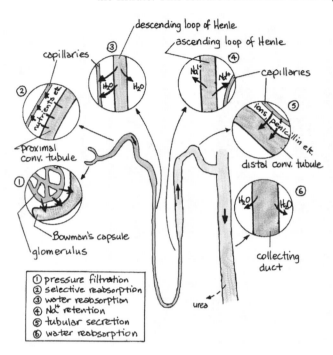

① pressure filtration
② selective reabsorption
③ water reabsorption
④ Na⁺ retention
⑤ tubular secretion
⑥ water reabsorption

Figure J-4. Urine Formation. *The filtrate that enters nephron tubules (1) is converted to urine by a series of events starting with the active reabsorption of nutrients (2). As the filtrate moves down the descending side of the loop of Henle (permeable to water, not to Na^{1+}), it is concentrated by water removal (3). As it moves back to the cortical region of the kidney, Na^{1+} is removed, first passively, then actively, (4) making the filtrate less concentrated once again, but this time with far less volume. In the distal convoluted tubule, the composition of blood is adjusted, a process known as tubular secretion (5). Finally, the urine makes its last pass through the medulla where more water is removed (6) making the smaller volume of urine more concentrated again.*

For obvious reasons, nephrons handle urea differently than glucose. Urea is a waste product that is toxic to the body in large concentrations. It has a very low threshold level in blood meaning that its concentration in plasma is supposed to be low. It enters the nephron and remains there becoming part of the urine. Glucose, on the other hand, has a higher threshold level in plasma therefore, after it enters the nephron, it is readily transported back into the blood. The reabsorption of glucose requires the presence of ATP and transport proteins along the walls of the tubules. Normally all the glucose is reabsorbed into the blood. When the blood passes through the liver, the hormone insulin promotes the removal of excess glucose, which is stored in the liver as glycogen. This mechanism maintains the low plasma concentration of glucose (about 0.1%).

The system malfunctions for diabetics who are unable to produce enough insulin. Therefore, their glucose concentration in plasma is higher than normal. It follows that less glucose is reabsorbed and some glucose remains in urine. The presence of the glucose in urine means that proportionately less water is reabsorbed. Abnormally large and frequent urine flows are symptomatic of diabetes. A **diuretic** is something that increases urine flows. For diabetics, insulin treatments are an **antidiuretic.**

REGULATING URINE FORMATION

The requirements of the body and the conditions under which the body works are not constant; therefore the functioning of the urinary system cannot be constant. **Antidiuretic hormone (ADH)** is secreted by the posterior pituitary gland in response to diminished blood volume. This hormone occupies receptor sites on the collecting ducts, which increase its permeability to water enabling more water to be reabsorbed into blood. In this manner, ADH counteracts low blood volume (and therefore pressure) by decreasing the volume of urine production. Alcohol is a diuretic because it inhibits ADH production thus promotes larger urine volumes. Coffee is also a diuretic, but for a different reason. It increases the blood volume by causing water retention. This increases pressure filtration and therefore promotes larger urine flows. Excessive consumption of diuretics can lead to **dehydration**.

Should blood pressure drop, perhaps due to low Na^{1+} concentration, resulting in less water being returned to plasma, the whole process could be in jeopardy. Conditions like this set off a chain reaction ending with the release of the hormone **aldosterone** from the adrenal cortex. Aldosterone increases the amount of Na^{1+} reabsorption at the distal convoluted tubule and thus increases the retention of Na^{1+} in plasma (which is coupled with an increase in the amount of K^{1+} excreted). The effect of this is the same as that of ADH - more water is reabsorbed and urine becomes more concentrated. This compensates for the initial low blood pressure.

CHECK YOUR UNDERSTANDING

MULTIPLE CHOICE:

1. Which process accounts for the return of needed nutrients to the blood?
   - A. Filtration.
   - B. Secretion.
   - C. Excretion.
   - D. Reabsorption.

2. Which part of a nephron is **MOST** affected by ADH?
   - A. Loop of Henle.
   - B. Collecting duct.
   - C. Distal convoluted tubule.
   - D. Proximal convoluted tubule.

3. Tubular secretion of histamines occurs in the
   - A. loop of Henle.
   - B. collecting duct.
   - C. distal convoluted tubule.
   - D. proximal convoluted tubule.

4. The pH of blood is increased by the movement of hydrogen ions from the
   - A. distal convoluted tubule into the peritubular capillaries.
   - B. peritubular capillaries into the distal convoluted tubule.
   - C. peritubular capillaries into the proximal convoluted tubule.
   - D. proximal convoluted tubule into the peritubular capillaries.

5. The cells of the loop of Henle are specialized for
   - A. filtration.
   - B. transport.
   - C. secretion.
   - D. contraction.

6. In which kidney region are the majority of Bowman's capsules found?
   - A. Renal vein.
   - B. Renal pelvis.
   - C. Renal cortex.
   - D. Renal medulla.

7. Which is **LEAST** likely to be found in the filtrate of a nephron?
   - A. Water.
   - B. Proteins.
   - C. Nutrients.
   - D. Nitrogenous wastes.

8. High solute concentration in the blood is detected by the
   - A. glomerulus.
   - B. distal tubule.
   - C. hypothalamus.
   - D. posterior pituitary gland.

9. Which **MOST** correctly describes where urea enters and leaves blood?
   - A. Entry at liver, removal at liver.
   - B. Entry at kidneys, removal at liver.
   - C. Entry at liver, removal at kidneys.
   - D. Entry at ileum, removal at kidneys.

10. Which would cause the kidneys to reabsorb more water?
    - A. Increased blood volume.
    - B. Increased cardiac output.
    - C. Decreased blood pressure.
    - D. Decreased ADH concentration.

11. When glucose is reabsorbed, it moves from the
    - A. blood to liver.
    - B. liver to blood.
    - C. filtrate to blood.
    - D. blood to filtrate.

12. If the pH of blood is too high
    - A. more H^{1+}, Na^{1+} and HCO_3^{1-} are excreted.
    - B. more H^{1+} is excreted, while less HCO_3^{1-} is reabsorbed.
    - C. fewer H^{1+} is excreted, while more HCO_3^{1-} is reabsorbed.
    - D. less H^{1+} and HCO_3^{1-} are reabsorbed.

13. Which would have the highest concentration of nitrogenous wastes?
    - A. Renal vein.
    - B. Renal artery.
    - C. Efferent arteriole.
    - D. Peritubular capillary bed.

14. Urea is a component of the fluid that passes from the
    - A. distal convoluted tubule into blood.
    - B. glomerulus into Bowman's capsule.
    - C. proximal convoluted tubule into blood.
    - D. peritubular capillaries into the loop of Henle.

15. Which is a correct function of a peritubular capillary bed?
    - A. Filtering of blood.
    - B. Reabsorption of nutrients.
    - C. Transport of blood to the liver.
    - D. Gathering wastes for excretion.

16. Deamination of amino acids is **MOST** directly responsible for
    - A. bile.
    - B. urea.
    - C. uric acid.
    - D. ammonia.

17. Which can be found in plasma and urine, but **NOT** in glomerular filtrate?
    - A. Urea.
    - B. Glucose.
    - C. Histamines.
    - D. Inorganic salts.

18. If alcohol intake increases,
    - A. ADH secretion increases.
    - B. ADH secretion decreases.
    - C. aldosterone secretion increases.
    - D. aldosterone secretion decreases.

19. The cells that line the proximal convoluted tubule are specialized for
    - A. peristalsis.
    - B. transmission of impulses.
    - C. active transport of nutrients.
    - D. intracellular digestion of nutrients.

20. Which is the pathway of glucose through the kidney of a healthy person?
    - A. Glomerulus – Bowman's capsule – loop of Henle – proximal tubule.
    - B. Renal vein – glomerulus – afferent arteriole – peritubular capillaries.
    - C. Afferent arteriole – Bowman's capsule – glomerulus – efferent arteriole.
    - D. Glomerulus – Bowman's capsule – proximal tubule – peritubular capillaries.

21. Increased blood volume would most likely result in
    - A. increased secretion of aldosterone and ADH.
    - B. decreased secretion of aldosterone and ADH.
    - C. decreased aldosterone secretion, but increased ADH secretion.
    - D. decreased ADH secretion, but increased aldosterone secretion.

22. Which structure is physically farthest away from all the others?
    - A. Ureter.
    - B. Distal tubule.
    - C. Loop of Henle.
    - D. Proximal tubule.

23. Where are the waste fluids moved by peristalsis?
    - A. Ureter.
    - B. Urethra.
    - C. Loop of Henle.
    - D. Collecting duct.

24. Which is **NOT** related to the release and function of aldosterone?
    - A. Secretion.
    - B. Excretion.
    - C. Absorption.
    - D. Elimination.

25. Which nephron part is physically closest to the renal pelvis?
    - A. Glomerulus.
    - B. Loop of Henle.
    - C. Distal convoluted tubule.
    - D. Proximal convoluted tubule.

WRITTEN ANSWERS:

1. State **ONE** function of each of the structures indicated in the diagram below.

    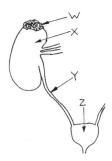

2. Below is a partially labeled diagram of a kidney.

    a. Name the labeled structures.
    b. List **TWO** differences between the fluids in A and B.
    c. List **TWO** differences between the fluids in B and C.
    d. Draw in and label the cortex, medulla and pelvic regions.
    e. Draw in and label the location of a nephron and collecting duct.
    f. What is the function of the renal pelvis?

3. Name the parts of the following diagram of a nephron and its blood vessels.

    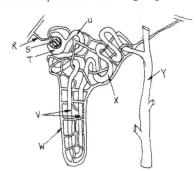

4. Refer specifically to the structure and function of the parts of a nephron to describe the details of the process of urine formation.

5. What is the role of the hypothalamus and posterior pituitary in urine formation?

UNIT K – REPRODUCTIVE SYSTEM

Male and female reproductive systems are homologous to each other. Both consist of **gonads** (gamete-producing organs), and tubes (or ducts) that conduct the gametes toward the outside. Each system, however, is specialized in its own way.

It is during fetal development that these specializations begin to materialize. The **testes**, (male gonads) are pulled to the outside through the abdominal wall to be housed in the **scrotum**. The route the testes follow becomes known as the **inguinal canal**. If the movement of the testes is incomplete, as with some male babies that are born prematurely, surgery may be required to release the testes from the body cavity or sterility can result. The temperature inside the body (37°C) is too high for the normal maturation of **sperm**. The female structures remain internal.

This unit individually examines the structures and functions of the male and female reproductive systems as well as considers the hormones and their control mechanisms. With the recognition that the systems are designed for reproduction, the unit includes a brief consideration of the changes that occur in the female system should fertilization occur. The unit closes with the topic of childbirth.

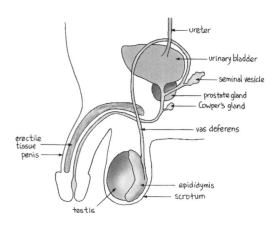

Figure K-1. Male Reproductive System. *The male reproductive system is mostly external where sperm can be produced at a temperature lower than 37°C. Mature sperm travel along the vas deferens to the penis, the organ of copulation.*

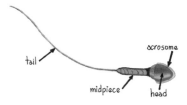

Figure K-2. Human Sperm Cell. *Sperm cells have three main sections: a head containing the genetic material, a mitochondria-rich midpiece that is associated with ATP generation, and the tail, which propels the sperm. A gel-like covering over their head called the acrosome contains enzymes to assist with penetrating the ovum.*

MALE REPRODUCTIVE SYSTEM

Male external genitalia consist of two structures: the scrotum containing the testes, and the **penis**. These are the beginning and the end of the route that sperm cells take through the male system. Sperm are produced in the **seminiferous tubules** of the testes by a process called **spermatogenesis**. Within the lining of these tubes is a particular layer of cells called the "germ layer". Sperm cells are derived from these unspecialized cells, which continuously divide by mitosis. One resulting daughter cell remains as the new germ cell; the other daughter cell moves towards the inner wall of the tubule and divides by meiosis to form cells, which eventually become sperm. Spermatogenesis begins in the seminiferous tubules but the developing sperm move through the **lumen** of the tubules to the **epididymis**.

Each testis has an epididymis on one side of its surface. The epididymis contains a coiled tube formed by seminiferous tubules after they join together. The tubes of the epididymis lead into the **ductus (vas) deferens**. The epididymis and the ascending portion of the vas deferens are the final maturation sites for sperm. It is in these regions of the tubules that sperm gain their swimming ability. The whole process takes a few days. Spermatogenesis begins during puberty and is ongoing well into old age. Millions of sperm cells develop every day throughout a male's reproductive life.

Mature sperm have three identifiable regions. The **head** contains the DNA (23 chromosomes). It is encapsulated by the **acrosome** a gel-like covering, which not only provides sperm with a "chemical guidance system" while swimming, but also with enzyme capabilities required for penetrating an **ovum** (egg). The **mid-piece** of a sperm contains mitochondria, which produce ATP required for the swimming activity. The **tail** (a flagellum) provides the locomotion.

Each testis also contains clusters of endocrine cells, called **interstitial cells** located between the seminiferous tubules. These cells secrete **androgens** (male sex hormones), including **testosterone**. Testosterone has many functions. It is essential for the maturation of the primary sex organs and development of the secondary sex characteristics associated with puberty. The secondary sex characteristics in males include ancillary hair growth, enlargement of the larynx, muscle and skeletal growth, release of oils and sweat (potentially causing acne and BO), and baldness (given appropriate genetics). Small amounts of androgens are also produced by the adrenal cortex contributing to muscle mass and strength in both males and females.

The penis is the organ of **copulation**. It contains **erectile tissue** and the urethra. This portion of the urethra serves a dual purpose. It not only conducts urine to the outside for excretion, but it also carries sperm during **ejaculation.** During sexual arousal, the arteries of the penis dilate (due to parasympathetic stimulation), and the veins become constricted allowing the erectile tissue to fill with blood making the penis erect. The changed angle of the penis allows sperm, bathed in **seminal fluid**, to enter the urethra. The seminal fluid is a milky white solution contributed to by three glands: **seminal vesicles**, **Cowper's (bulbourethral) gland**, and the **prostate gland**. Semen has a pH of 7.5, contains fructose and citrate to nourish the sperm, HCO_3^{1-} as a buffer, and a lipid-based chemical called **prostaglandins**. Prostaglandins promote uterine contractions, which aid sperm activity after intercourse.

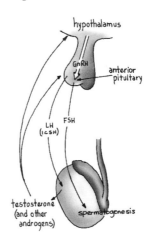

When **semen** (sperm plus seminal fluid) enters the urethra it promotes the rhythmic contractions leading to ejaculation. A sphincter at the base of the bladder is closed during this time, so that urine does not enter the urethra. After ejaculation, the penis returns to its **flaccid** state for the duration of a **refractory period**.

With puberty, the gonads are stimulated into maturation by **gonadotropic hormones**, which are released from the anterior pituitary gland. The release of these hormones is triggered by the hypothalamus through its release of **gonadotropic releasing hormones (GnRHs)** to the anterior pituitary. The anterior pituitary then responds by releasing two gonadotropic hormones: **luteinizing hormone (LH)** (a.k.a. **interstitial cell stimulating hormone, ICSH**) and **follicle stimulating hormone (FSH)**. LH promotes the production of testosterone from the interstitial cells, where FSH promotes the maturation of sperm. Testosterone is maintained at a fairly constant level by negative feedback because increased levels of testosterone inhibit the hypothalamus and the pituitary from releasing their hormones.

Figure K-3. Feedback Control of Testosterone Levels. *Testosterone release from the interstitial cells of the testes is promoted by luteinizing hormone (LH) from the anterior pituitary. Testosterone levels decrease the amount of LH. In this manner, the levels of testosterone are maintained. This is an example of negative feedback.*

Table K-1. Components of Seminal Fluid

Source Gland	Contribution to Seminal Fluid
Seminal vesicles	Mucus, fructose, and prostaglandins
Prostate gland	Fluid, citrate
Cowper's gland	Fluid, HCO_3^{1-}

K.1 Concept Check-up Questions:
1. Name the structures along the route that sperm travel from where they are produced to where they leave the male body.
2. a. How do the brain and anterior pituitary work together to trigger the onset of puberty?
   b. What are the effects of testosterone release?
3. Where does each of the following components of semen originate: sperm, nutrients, fluid medium and buffers?
4. What are the roles of the epididymis and the acrosome?

FEMALE REPRODUCTIVE SYSTEM

Even though male and female reproductive structures are homologous in their design, they appear very different from each other. The **ovaries** are the structures where **ova** (singular, **ovum**) are produced. The ovaries lie next to the beginning of the female reproductive tract, which consists of the **oviducts**, **uterus**, and **vagina**. This tract does not meet up with the urethra as it does in males.

Oogenesis is the development of ova. At birth, a female has several hundred thousand immature ova in structures called follicles located in the cortical regions of her ovaries. After puberty, these mature, roughly one a month, over her reproductive life. Only a few hundred ever mature. As a follicle matures, it undergoes changes to house the ovum in a fluid filled space. This development continues until the structure bulges out from the ovary surface. At this point the

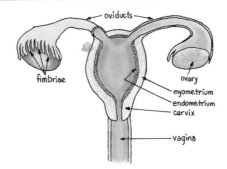

Figure K-4. Female Reproductive System. *The female reproductive system consists of two ovaries, which produce ova. The ova travel to the uterus through the oviducts and pass to the outside through the cervix and vagina. In the event of fertilization, which generally occurs in the oviducts, the egg may become implanted in the uterus. This marks the onset of pregnancy.*

follicle is mature and is known as a **Graafian follicle**. **Ovulation**, the bursting of the follicle and the release of the ovum, is under hormonal control. The ovum, once freed, is directed into the oviduct by the undulations of the **fimbriae** that surround the opening of the oviduct. The oviduct is a muscular tube that is lined with cilia. The cilia sweep fluid towards the uterus. In this manner, the ovum, without its own means of locomotion moves along the female tract to the uterus. This takes four or five days. The remaining follicular structure in the ovary develops into a glandular tissue known as the **corpus luteum**.

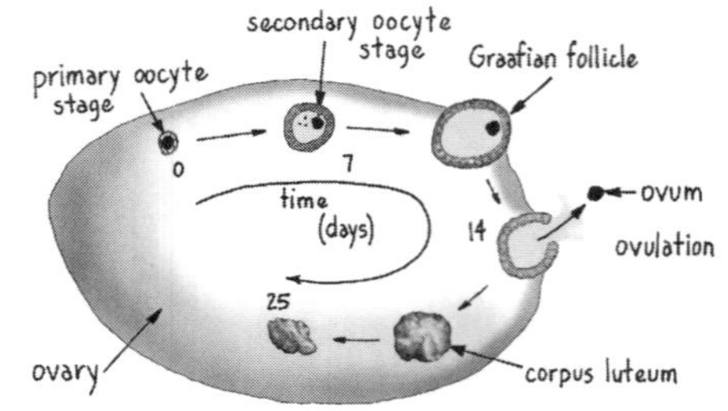

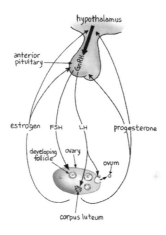

Figure K-5 Oogenesis. *It takes about two weeks for an ovum to mature in an ovary. When mature, the ovum is contained in a follicle, which bursts (ovulation) to release the ovum. The remaining tissue specializes as a temporary endocrine gland called the corpus luteum. This gland degenerates after while and the cycle starts over again with a new follicle.*

The uterus is sometimes called the **womb**. It is a thick-walled muscular pear-shaped organ. The outer muscular layer is called the **myometrium**. Its inner lining is called the **endometrium**. The endometrium itself has two layers, a **basal layer**, which is adjacent to the muscle and elastic fibres of the myometrium, and a functional inner lining. The thickness of the inner lining varies according to the hormone levels of the reproductive cycle.

The vagina is the tube that extends from the **cervix** (neck of the uterus) to the outside. Its inner lining consists of **mucosal cells**. It serves as the **birth canal** and the female organ of copulation. The **clitoris** is the female structure that is homologous to the penis in males. It is located at the anterior end of the **labia** (folds of skin) that surround the vaginal opening. Where the male penis becomes engorged with blood during sexual arousal, the clitoris, labia and vaginal walls of the female become similarly engorged.

Much like the male system, the female system is hormonally regulated. FSH stimulates the maturation of an ovum in the ovary. LH stimulates the secretion of the female hormones **progesterone** and **estrogen** from other cells of the follicle. Progesterone stimulates the maturation of the endometrium lining of the uterus. Estrogen is a more complex hormone. It plays a role in the maturation of ova as well as the development of secondary sex characteristics in females (uterine cycle, fat and hair distribution, growth of the uterus and vagina, pelvic growth, and it contributes, along with progesterone, to breast development).

Figure K-6. Feedback Control in Females. *Unlike hormone levels in males, the amount of female hormones released varies over the 28-day period of time, which result in the cyclical changes in the ovaries and the uterus. Their levels are generally controlled by negative feedback; however, a high estrogen level causes a sudden flood of LH and (to a lesser extent) FSH, which promotes ovulation.*

As the secretion of estrogen increases, it seems to have a negative feedback relationship on the anterior pituitary because the rate of release of LH and FSH begins to decrease. However, when the follicle is mature and the level of estrogen is the highest (a process that takes about fourteen days), there is a sharp increase in the concentrations of LH and, to a lesser extent, FSH. It is believed that this significantly high level of estrogen exerts *positive* feedback on the hypothalamus, thus causing a surge of LH (and FSH) resulting in ovulation.

After ovulation, the corpus luteum (the follicular tissue that remains in the ovary) temporarily retains an endocrine function. During this time, it continues to secrete progesterone and, after a few days, renews its role of secreting estrogen. Estrogen stimulates the continued thickening of the endometrium. Progesterone stimulates the endometrium to mature and become secretory. These changes are in preparation for the possibility of fertilization and pregnancy. Provided that fertilization and pregnancy do not occur, the corpus luteum degenerates after about ten days. When it degenerates, the levels of estrogen and progesterone drop suddenly and the endometrium that was built up and maintained to provide for an embryo is shed. The shedding of the endometrium (mostly blood tissue) is known as **menstruation (menses)**. The period of menses typically lasts about five days. During menses, the anterior pituitary (once again) begins to release LH and FSH and the cycles of change that occur in the ovaries and uterus begin anew.

Table K-2. Ovarian and Uterine Cycles

Ovarian Cycle		Uterine Cycle	
Phase (Time)	**Events**	**Phase (Time)**	**Events**
Follicular Phase (Day 1 to 13)	LH and FSH secretion by the anterior pituitary gland Maturation of a follicle and secretion of estrogen (and progesterone) to build the endometrium	**Menstrual Phase** (Day 1 - 5)	Endometrium lining breaks down and is discharged = **Menses**
		Proliferative Phase (Day 6 - 14)	Rebuilding and thickening of the endometrium
Ovulation (Day 14)	The ovary releases an ovum. Corpus luteum is formed.		
Luteal Phase (Day 15 - 28)	Estrogen secretion by the corpus luteum causing the endometrium to continue to thicken and mature. Progesterone secretion is making the endometrium secretory.	**Secretory Phase** (Day 15 - 28)	Endometrium continues to thicken. Maturation of the secretory glands of the endometrium

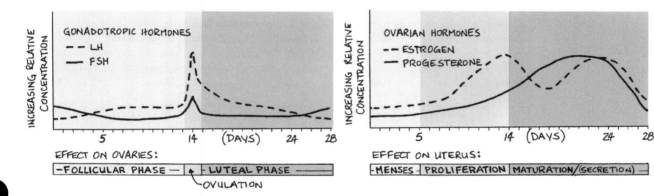

Figure K - 7. Cyclical Hormonal Changes. *The level of gonadotropic and uterine hormones varies over a 28-day cycle. The gonadotropic hormones cause changes in the ovaries such as follicle development, ovulation, and the promotion of the ovarian hormones. These hormones, in turn, affect the uterus in order to prepare it in the event of a pregnancy.*

K.2 Concept Check-up Questions:
1. a. Describe the physical events of the phases of the ovarian cycle.
   b. What is a Graafian follicle?
   c. What is the purpose of the corpus luteum?
2. a. Assuming that fertilization has not occurred, describe the route that ova take through the female body.
   b. By what mechanisms are they transported?
3. a. Describe the physical events of the phases of the uterine cycle.
   b. How do the ovarian hormones affect the uterine cycle?
4. Describe one example of negative feedback and one example of positive feedback in the female reproductive system.

IMPLANTATION AND PREGNANCY

During copulation semen, containing as many as 400 million sperm, is **ejaculated** into the vagina. The sperm, guided by their acrosomes, begin swimming towards the ovaries. An ovum is only viable for fertilization for six to twenty-four hours after ovulation. As a result, fertilization usually occurs in the oviduct (recall that it takes a few days for an ovum to reach the uterus). The resulting zygote begins to divide by mitosis as the structure continues towards the uterus where it embeds (implants) itself in the endometrium wall. The zygote has become a multi-celled structure by this time – the beginning of an embryo. With **implantation**, starts the production of a new hormone, **HCG (human chorionic gonadotropin)**, which prevents the corpus luteum from degenerating. This marks the beginning of **pregnancy**. Pregnancy tests test for the presence of HCG in urine. A new tissue results from the combination of embryonic and maternal tissues following implantation. This tissue forms the **placenta**, which is an organ designed for the exchange of plasma components between the fetus (embryo) and the mother. The placenta is primarily an interface between the two circulatory systems.

The maintenance of the corpus luteum by HCG means that its production of progesterone and estrogen continues. In time, the **placenta** fully develops and begins to secrete estrogen and progesterone. The progesterone maintains the endometrium and prevents menstruation from occurring. The estrogen inhibits LH and FSH release from the anterior pituitary. By the tenth week of a pregnancy, the placenta is developed enough to produce the quantities of estrogen and progesterone required for maintaining the pregnancy. The secretions of HCG diminish and the corpus luteum finally degenerates. The pregnancy is well established and the fetus will continue to grow and develop.

The muscular uterus continues to contract slightly throughout a pregnancy. Near the end of the pregnancy, however, these contractions become more severe (**false labour**). True labour begins when the contractions become sustained. The birth process is considered to have three stages: dilation of the cervix, birth, and expulsion of the afterbirth. **Oxytocin** is a hormone that is released by the posterior pituitary. Its release is stimulated by pressure on the cervix, which is a direct result of the contractions of the uterus. Oxytocin causes further contractions of the uterus, therefore more pressure, which increases oxytocin release and so on. The relationship between oxytocin and the uterus is an example of **positive feedback** (where a change reinforces itself). The intense uterine contractions combined with the eventual dilation of the cervix leads to the birth of the child. Expulsion of the afterbirth (placenta and umbilical cord) follows.

K.3 Concept Check-up Questions:
1. What physical events mark the beginning of pregnancy? Why do they occur?
2. Describe the origin, role, and fate of a placenta.
3. How is a corpus luteum affected by pregnancy?
4. What causes labor pains associated with childbirth?

CHECK YOUR UNDERSTANDING

MULTIPLE CHOICE:

1. Which gland is an endocrine gland?
   A. Testis.
   B. Prostate gland.
   C. Cowper's gland.
   D. Seminal vesicle.

2. Which hormone is **MOST** associated with childbirth?
   A. estrogen.
   B. oxytocin.
   C. progesterone.
   D. chorionic gonadotropin.

3. Chorionic gonadotropin is different from other gonadotropins because it does **NOT**
   A. act on the gonads.
   B. enter the bloodstream.
   C. stimulate any tissue in the body.
   D. come from the anterior pituitary.

4. In males, the tube that conducts both semen and urine is the
   A. ureter.
   B. urethra.
   C. vas deferens.
   D. common bile duct.

5. Sperm do **NOT** pass through the
   A. epididymis.
   B. vas deferens.
   C. seminal vesicles.
   D. seminiferous tubules.

6. Which hormone **DIRECTLY** causes ovulation?
   A. Estrogen.
   B. Oxytocin.
   C. Progesterone.
   D. Luteinizing hormone.

7. What occurs during ovulation?
   A. An egg cell matures.
   B. A zygote implants in the uterus.
   C. Sperm are ejected from the penis.
   D. An egg is released from the ovary.

8. Implantation normally occurs
   A. during fertilization.
   B. a few hours after fertilization.
   C. a few days after fertilization.
   D. a few weeks after fertilization.

9. After ovulation, the ruptured follicle
   A. disappears as its cells degenerate.
   B. leaves the body as waste material.
   C. becomes a temporary endocrine gland.
   D. becomes part of the ovary's epithelium.

10. Which of these is a male **PRIMARY** sex characteristic?
    A. Deepening of the voice.
    B. Facial and pubic hair growth.
    C. External reproductive organs.
    D. Muscle and skeletal development.

11. The testes are suspended in a scrotum because
    A. that is where the penis is located.
    B. there is no room inside the body.
    C. this arrangement assists sperm and semen to flow into the urethra.
    D. spermatogenesis occurs below 37 °C.

12. Sperm cells are stored mainly in the
    A. urethra.
    B. prostate.
    C. epididymis.
    D. seminal vesicles.

13. Correctly complete the following statement. "Sperm cells..."
    A. carry their own nutrients.
    B. are less numerous than ova.
    C. are each individually motile.
    D. always swim towards the penis.

14. Which does **NOT** contribute to the fluids that make up seminal fluid?
    A. Testes.
    B. Prostate.
    C. Cowper's glands.
    D. Seminal vesicles.

15. Testosterone is produced by
    A. sperm cells.
    B. hypothalamus.
    C. interstitial cells.
    D. anterior pituitary.

16. Which set of structures becomes engorged with blood during sexual responses?
    A. vagina, labia, breasts, penis.
    B. clitoris, labia, vagina, penis.
    C. clitoris, labia, breasts, penis.
    D. clitoris, vagina, breasts, penis.

17. Select the true statement.
    A. The rate of spermatogenesis is much more variable than oogenesis.
    B. Spermatogenesis begins before birth and oogenesis begins at puberty.
    C. Estrogen promotes oogenesis and progesterone promotes spermatogeneis.
    D. Oogenesis produces one functional gamete where spermatogenesis produces more.

18. Less GnRH released from the hypothalamus will
    A. increase LH and FSH secretion.
    B. decrease the release of gonatotropic hormones.
    C. stimulate an increase in progesterone and estrogen levels.
    D. initiate ovulation in females and reduce spermatogenesis in males.

19. Which hormone **INITIATES** endometrial development?
    A. LH
    B. FSH
    C. estrogen.
    D. progesterone.

20. The correct order of structures through which an unfertilized ovum passes is
    A. oviduct – uterus – cervix – vagina.
    B. uterus – vagina – cervix – oviduct.
    C. oviduct – vagina – uterus – cervix.
    D. oviduct – cervix – uterus – vagina.

21. Fertilization of ova usually occurs in the
    A. ovary.
    B. uterus.
    C. vagina.
    D. oviduct.

22. The three phases of the ovarian cycle are
    A. embryo, fetus, and newborn.
    B. follicular, ovulatory, and luteal.
    C. first, second, and third trimester.
    D. menstrual, proliferative, and secretory.

23. Progesterone is needed for
    A. uterine contractions.
    B. corpus luteum development.
    C. maturation of the endometrium.
    D. female secondary sex characteristics.

24. Which of the following has the **GREATEST** effect on the expression of female secondary sex characteristics?
    A. Estrogen.
    B. Progesterone.
    C. Luteinizing hormone.
    D. Follicle stimulating hormone.

25. Which of the following **BEST** describes how an ovum enters an oviduct?
    A. It swims.
    B. Peristalsis.
    C. It is drawn in by fluid currents.
    D. It is propelled by the force of the bursting follicle.

26. The main source of progesterone during the secretory phase is
    A. adrenal gland.
    B. corpus luteum.
    C. anterior pituitary.
    D. developing follicle.

27. The secretory phase of uterine cycle
    A. results from low LH and FSH.
    B. is when the menstrual flow occurs.
    C. is associated with dropping progesterone levels.
    D. corresponds to the luteal phase of the ovarian cycle.

28. The sources of the hormones that regulate the ovarian and uterine cycles are
    A. uterus and ovaries.
    B. anterior pituitary and ovaries.
    C. anterior pituitary, ovaries, and uterus.
    D. anterior and posterior pituitary and ovaries.

29. Menstrual discharge comes from the
    A. cervix.
    B. vagina.
    C. oviduct.
    D. endometrium.

30. Progesterone is produced in the
    A. ovary and acts on the testes.
    B. ovary and acts on the uterus.
    C. ovary and acts on the pituitary.
    D. pituitary and acts on the ovary.

31. Which is an **INCORRECT** match of structure with function?
    A. epididymis – stores sperm.
    B. scrotum – suspends the testes.
    C. ovary – secretes estrogen and progesterone.
    D. seminiferous tubules – produce seminal fluid.

32. The embryonic hormone that maintains the corpus luteum during pregnancy is
    A. Luteinizing hormone.
    B. Follicle stimulating hormone.
    C. Gonadotropic releasing factor.
    D. Human chorionic gonadotropin.

33. Through which structures must a child pass during childbirth?
    A. Cervix and vagina.
    B. Vagina and urethra.
    C. Fallopian tube and navel.
    D. Fallopian tube and umbilical cord.

34. HCG in a woman's urine indicates that
    A. she has just ovulated.
    B. menstruation is about to begin.
    C. her levels of estrogen are decreasing.
    D. implantation of an embryo has occurred.

WRITTEN ANSWERS:

1. Design a data table to contrast sperm and ova in **FOUR** ways. Complete your table.

2. Name and describe **ONE** function of each of the structures indicated in the diagram below.

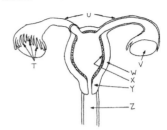

3. Four blood samples (A, B, C, and D) from the same woman were analyzed for their relative hormone levels on Day 1, 8, 13 and 23 of her menstrual cycle. The data (in random order) is presented in the chart below.

Hormone	A	B	C	D
FSH	Med	High	Low	Low
LH	Med	High	Low	Low
Progesterone	Med	Med	Low	High
Estrogen	Med	High	Low	High

Use this data to determine the **MOST LIKELY** day each of the samples was taken.

4. Examine the diagram of an ovary below and answer the associated questions.

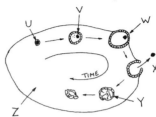

a. Name the labeled structures.
b. What event is occurring at **X**?
c. What is the role of structure **Y**?
d. Describe the condition of the endometrium at the same time as **X** occurs.

5. Examine the diagram of a testis below. Name and describe the function of each of the labeled structures.

6. a. Name **ONE** role of each of the following in relation to a female's reproductive system.
   - hypothalamus

   - anterior pituitary

   - posterior pituitary

   b. State **TWO** ways the neuroendocrine function is different between males and females.

7. What are the components of semen? Why is each one important?

8. List **SEVEN** hormones associated with the reproductive system. and identify the source, target organ, and effect of each.

9. What is a GnRH? What does a GnRH do?

10. Compare and contrast each of the following:
    a. ejaculation and ovulation

    b. follicle and corpus luteum

11. Describe the reproductive roles of the hypothalamus and pituitary glands.

APPENDIX I - EXPERIMENTAL DESIGN

Science is more than information. It also encompasses a set of processes that enable scientific endeavors to determine new information. The set of processes that is used has become known as the **scientific method**.

This method, or way of doing things, starts with making **observations**. Observations can be made visually or by recording events. Observations can be **qualitative** (describing quality) or **quantitative** (describing quantity). An analysis of observations can lead to a **hypothesis**, or an educated guess that will account for the observations. A hypothesis will also allow one to make a **prediction** and design an **experiment** with **procedures** that will test the accuracy of the prediction.

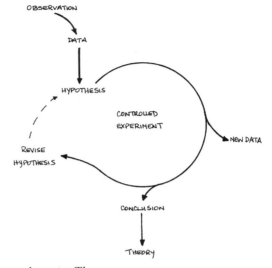

During an experiment, additional data is collected. If the data supports the hypothesis, a **conclusion** can be made. If the same conclusion is arrived at through many independent experiments, a **theory** can be devised. A theory is a hypothesis that had withstood testing. If the data from an experiment does not support the hypothesis, then the hypothesis is called a **null hypothesis** and it can be discarded or revised and the experiment can be redesigned to accommodate the revised hypothesis.

In order that scientists and experimenters can be certain that the results of their research can be trusted (reliable), they have to make sure that if the experiment were to be repeated, the results would come out the same. Valid experiments are those that are based on, or supported by evidence, and therefore, not open to interpretation. **Reliability** and **validity** are achieved through the use of **controlled experiments**. These experiments provide a standard of comparison between their components. The part of the experiment that provides this standard in called the **control**.

In the typical design of a controlled experiment, a large sample of test "subjects" is used. The advantage of a large **sample size** is so results brought about by individual differences are minimized. The subjects are randomly divided into an **experimental group** and a **control group** (ideally of equal size). Both groups receive the exact same treatment, with the exception of one **variable**. The experiment is designed to test the effects of this variable on the subjects in the experimental group. Any difference in results between the two groups can be attributed to the independent variable being tested. For example, in an experiment to test the effects of nutrient fertilizer on plants, any difference in growth (**dependent variable**) would be a result of the fertilizer (**independent variable**) as long as all the other variables were controlled (kept the same) and the sample sizes were large enough.

A full application of all of these concepts is in the following example. Suppose a person (let's call her Maggie) noticed that unfertilized sea urchin eggs were not always the same size. They were always close, but not exactly the same. These cells had identical functions (unfertilized eggs) and because cell size affects function, there had to be some explanation for the noted differences in size. Maggie figured that the size of the eggs was due to the salt concentration in the water. Sea urchins are marine invertebrates, and there is no guarantee that the concentration of the salt in different parts of the ocean is constant. For example, it could be lesser concentrated near the mouth of a stream or river. This line of reasoning allowed Maggie to form the hypothesis that sea urchin eggs collected near the mouth of a river would be larger than those collected at a great distance from the source of any fresh water. This hypothesis is based on the observation that cells swell when placed in hypotonic environments. Maggie predicted that this would be true for sea urchin eggs.

Maggie realized it would be impractical to travel around to different parts of the ocean collecting unfertilized sea urchin eggs, so she designed an experiment to simulate real conditions. In the lab, she collected a large sample of sea urchin eggs and divided them up into two beakers of seawater with a salinity of 3.5%. She called one beaker Group A left it alone to be used as the control group She added fresh water to the other beaker to dilute the salt concentration to 3.0%, which she called group B, the experimental group. She then set about to microscopically measure the diameter of the eggs. The diameters of many eggs in both the control group and the experimental group would be averaged out. If her hypothesis is correct, she should find that the average diameter of the eggs in the experimental group would be larger than that of the control group.

In this experimental design, the independent variable is the salt concentration. The dependent variable is the diameter of the eggs. It follows *that the diameter of the eggs depends on the salt concentration.*

Experiments like this are often expanded into ones where the effect of a range of conditions is determined. There would be nothing to stop Maggie from setting up an experiment with several different beakers where the salinity increased incrementally by 0.5% from 2.0% to 6.0%. Suppose she did this and collected the following data:

Beaker	A	B	C	D	E	F	G	H	I
Salinity	2.0%	2.5%	3.0%	3.5%	4.0%	4.5%	5.0%	5.5%	6.0%
Average Diameter (μm)*	10.8	10.5	9.9	9.2	8.4	7.7	7.4	7.3	7.3

*Hypothetical data

Interpretation of these results can follow. These are possible questions and answers:

1. **Does it support the hypothesis?**

   The answer is yes. The hypothesis was that sea urchin eggs in water of lower salinity would be larger.

2. **Account for this observation.**

   When placed in a hypotonic environment, water will move by osmosis into the egg, which swells as a result.

3. **Graph the results.**

   To graph these results, the independent variable (salinity) is put along the horizontal x-axis. The dependent variable (average diameter) is put on the vertical y-axis. The points are then plotted and a line drawn to connect the points.

Graph App. 1. Diameter of Sea Urchin Eggs vs. Salinity

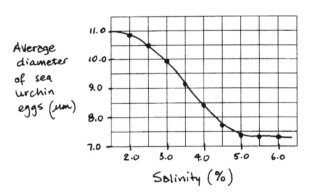

4. **Explain the shape of the graph.**

   At lower environmental salt concentrations, water enters the eggs. This increases turgidity. There is a limit to the amount of water that the cells will take in, which is reflected by the fact that the graphed line levels out at the lower concentrations. At higher salt concentrations (hypertonic), water leaves the cells and they become dehydrated. The cells only contain so much water, a fact that is observable as the line levels out at the higher concentrations. Somewhere in between these two extremes, the sea urchin eggs would be isotonic to the environmental salt water.

The topic of experimental design can be applied to a few units in Biology 12. The above application is to the transport topic, where osmosis creates a measurable difference in cell size. Another one would be the effect of an environmental condition (like pH or temperature) on enzymatic activity. Enzymatic activity can be determined as the rate of production of a product, or the rate of consumption of a substrate. It would be good practice to use the theory of the enzyme unit to the topic of digestion and attempt to design an experiment. For example, design an experiment to determine the optimum pH for trypsin activity.

Appendix II   Genetic Code

First Nucleotide in the Codon				
U	UUU phenylalanine (phe) UUC phenylalanine (phe) UUA leucine (leu) UUG leucine (leu)	UCU serine (ser) UCC serine (ser) UCA serine (ser) UCG serine (ser)	UAU tyrosine (tyr) UAC tyrosine (tyr) UAA **STOP** UAG **STOP**	UGU cysteine (cys) UGC cysteine (cys) UGA **STOP** UGG tryptophan (trp)
C	CUU leucine (leu) CUC leucine (leu) CUA leucine (leu) CUG leucine (leu)	CCU proline (pro) CCC proline (pro) CCA proline (pro) CCG proline (pro)	CAU histidine (his) CAC histidine (his) CAA glutamine (gln) CAG glutamine (gln)	CGU arginine (arg) CGC arginine (arg) CGA arginine (arg) CGG arginine (arg)
A	AUU isoleucine (ile) AUC isoleucine (ile) AUA isoleucine (ile) AUG methionine (met)	ACU threonine (thr) ACC threonine (thr) ACA threonine (thr) ACG threonine (thr)	AAU asparagine (asn) AAC asparagine (asn) AAA lysine (lys) AAG lysine (lys)	AGU serine (ser) AGC serine (ser) AGA arginine (arg) AGG arginine (arg)
G	GUU valine (val) GUC valine (val) GUA valine (val) GUG valine (val)	GCU alanine (ala) GCC alanine (ala) GCA alanine (ala) GCG alanine (ala)	GAU aspartate (asp) GAC aspartate (asp) GAA glutamate (glu) GAG glutamate (glu)	GGU glycine (gly) GGC glycine (gly) GGA glycine (gly) GGG glycine (gly)
Second Nucleotide in the Codon	**U**	**C**	**A**	**G**

Each of the 64 possible codons is organized into columns according to the first and then the second nucleotide in the sequence. 61 of the codons code for a specific amino acid, which is identified by name and abbreviation (in brackets). The three codons that do not code for an amino acid are called **STOP** codons. Their incorporation into a mRNA strand signals the termination of the polypeptide. A single codon, AUG, which codes for methionine, serves as the **START** codon.

Appendix III   Hormones

Hormone	Source	Target Organ	Release Triggered by	Effect
ADH	posterior pituitary	collecting ducts	low blood volume; high [solute] in plasma	increases water reabsorption from urine
adrenalin	adrenal medulla	various organs	high concentration of noradrenalin	stimulates fight/flight responses
aldosterone	adrenal cortex	distal convoluted tubules	low [Na^{1+}] in plasma; low blood pressure / volume	increases sodium retention (and potassium excretion)
estrogen	follicle cells	2° sex structures	LH	promotes 2° sex characteristics causes LH and FSH surge
	corpus luteum	uterus	FSH	building of the endometrium during the follicular phase
FSH	anterior pituitary	gonads	GnRF from hypothalamus	promotes oogenesis (♀) and spermatogenesis (♂)
glucagon	pancreas	liver	low blood sugar	promotes release of glucose from glycogen storage
GnRH	hypothalamus	anterior pituitary	plasma hormone levels	promotes the release of LH (ICSH) and FSH
HCG(H)	uterine walls; developing placenta	ovary	implantation	maintains the corpus luteum thus ensuring adequate levels of progesterone
insulin	pancreas	liver	high blood sugar	increases conversion of glucose to glycogen
LH (♀)	anterior pituitary	ovary	GnRH from hypothalamus	stimulates release of estrogen and progesterone
			high level of estrogen	causes ovulation (positive feedback) and promotes development of corpus luteum after ovulation
LH (ICSH) (♂)	anterior pituitary	interstitial cells	GnRH from hypothalamus	promotes testosterone release
oxytocin	posterior pituitary	uterus	uterine contractions	stimulates uterine contractions (positive feedback)
progesterone	follicle cells	endometrium	FSH	stimulates the maturation and maintenance of the endometrium
	corpus luteum		LH	
prostaglandins	various tissues	uterus	(component of semen)	stimulate uterine contractions following intercourse
testosterone	interstitial cells of the testes (and adrenal cortex)	various organs	level maintained by ICSH release from the anterior pituitary	stimulates and maintains 2° characteristics in ♂ contributes to muscle development in ♂ and ♀
thyroxin	thyroid gland	body cells	negative feedback from the body	promotes cellular respiration, (increasing O$_2$ consumption and CO$_2$ production)

SECTION III - ANSWERS

UNIT A - THE CHEMISTRY OF LIFE

CONCEPT CHECK-UP QUESTIONS:

A.1
1. An organic molecule is one that contains carbon.
2. A polar covalent bond is a bond between atoms that is formed by the unequal sharing of electrons.
3. H-bonding is caused by the attraction between opposite dipoles. H-bonding results in cohesion among water molecules.
4. Water's solvation properties are important to life because organisms contain a lot of water and the water transports dissolved particles. Blood plasma, for example, is mostly water.

A.2
1. An acid releases H^{1+} when in solution; a base releases OH^{1-} (which reduces the amount of free H^{1+}) when in solution.
2. Water is neutral because it dissociates to release H^{1+} and OH^{1-} in equal numbers therefore the pH remains at 7.0 (neutral).
3. A solution with a low pH is more acidic than one with a high pH.
4. Bicarbonate ions are buffers in blood. They resist pH changes.

A.3
1. Dehydration synthesis is the building of larger molecules from monomers involving the release of water. Hydrolysis is the opposite, namely the degradation of polymers into their components requiring the addition of water.
2. Lipids do not form true polymers because there is a limit to the number of monomers of lipids that can be combined together.
3. When monomers are bonded together, the bond forms between an atom that has a hydrogen atom attached to it and one that has a hydroxide group attached to it. When the bond forms, these attachments must be released. Their combination forms water.
4. Glucose is made in photosynthetic plant cells and used either for energy (by mitochondria), stored as starch, or used to build structural molecules like cellulose, which is used to make cell walls. In animals, glucose is a product of digestion and used for energy or stored in liver tissue in the form of glycogen.

A.4
1. Lipids are molecules that do not mix with water. They are essentially non-polar molecules, where water molecules are polar.

2. Fatty acids from animal tissues tend to be (more) saturated and therefore more solid at room temperature. Fatty acids from plant tissues, such as vegetable oils, are unsaturated and less solid.
3. Glycerides can vary in terms of the number and placement of the fatty acids on the glycerol backbone, the length of the fatty acids, and the nature of the fatty acids (saturated vs. unsaturated).
4. Steroids and waxes are lipids that do not contain glycerol.

A.5
1. Amino acids have an amino group (terminus, or end) and a carboxylic acid group, both of which are attached to a central carbon, creating the N-C-C backbone. To the central carbon atom are attached a hydrogen atom and an "R" group.
2. A peptide bond is a strong covalent bond. Peptide bonds join amino acids together.
3. The two conformations (shapes) of secondary structure in proteins are the α-helix and the β-pleated sheet. Each is caused by a difference in the way that H-bonding occurs between nearby amino acids during the formation of the protein.
4. The hydrogen bonds that help maintain tertiary structure are the weakest in the molecular structure and, therefore break first when conditions are altered. This results in denaturation.

A.6
1. DNA nucleotides contain the sugar, deoxyribose; where RNA nucleotides contain ribose. As well, the base thymine does not exist in RNA. It is replaced by uracil, a very similar pyrimidine.
2. Size is the most obvious difference. Purines are larger, composed of two ring structures; Pyrimidines are single-ring structures.
3. Chromosomes are comprised of DNA and small protein complexes, known as histones.
4. The P~P bonds in ATP are high-energy bonds, but they are weak bonds, meaning they can be formed and broken readily. When they are broken, releasing P and generating ADP, the bond energy is freed and can be used by the cell. When additional energy is available to the cell, ATP can be regenerated for future use.

MULTIPLE CHOICE QUESTIONS:

1. C	6. D	11. B	16. B	21. D	26. A	31. D	36. C	41. B
2. D	7. D	12. C	17. C	22. B	27. C	32. B	37. C	42. C
3. D	8. A	13. B	18. C	23. C	28. C	33. A	38. B	43. A
4. D	9. C	14. A	19. D	24. B	29. B	34. C	39. C	
5. D	10. C	15. B	20. D	25. C	30. A	35. B	40. D	

WRITTEN ANSWER QUESTIONS:

1. a. A water molecule consists of one atom of oxygen and two atoms of hydrogen bonded together by polar covalent bonds. Oxygen is the central atom. The molecule is bent, and the bonding electrons are shared unequally – used primarily by oxygen, which gives oxygen a negative dipole and leaves the hydrogen atoms with positive dipoles.
   b. The polarity of the molecules allows them to form hydrogen bonds with other water molecules. The presence of the hydrogen bonds and the fact that water is so abundant provides the unique properties.
   c. Water is significant in cells because it:
      a) has a high specific heat capacity, which means that takes a lot of energy to change the temperature of water. This is important for organisms because chemical reactions that are part of metabolism are temperature sensitive.
      b) is cohesive with other water molecules and, as such, moves through blood vessels as a continuous fluid transporting nutrients and cells around the body.
      c) Is a solvent for inorganic substances, therefore it transports ions as well as (non-polar) molecules that will not dissolve in it.

2. Acids and bases differ primarily in the amount of free hydrogen ions that are in their respective solutions. If there is more free hydrogen ions than OH^{1-} then the solution is called acidic (acids release hydrogen). If there are less free hydrogen ions, then the solution is called basic (alkaline). Bases often release ions, which combine with hydrogen, thus reducing the amount of hydrogen in the solution. Increasing the amount of hydrogen lowers the pH. pH is the negative logarithm of the hydrogen ion concentration.

3. Two glucose molecules can be combined (using the enzyme, maltase) to form a molecule of maltose and water as a byproduct. This reaction is synthesis because a larger molecule is being made. The hydrolysis of maltose requires maltase again as well as the addition of water. When the enzyme cleaves maltose, the water is required to complete the bonds of the individual glucose molecules that are produced.

4. a. The unit molecules of: carbohydrates are monosaccharides (usually glucose); lipids are glycerol and fatty acids; proteins are amino acids and nucleic acids are nucleotides.
    b. Carbohydrates metabolized for energy. Lipids can be stored for energy reserves or used for insulation against damage and temperature loss, lubrication (as in ear wax, skin oils etc.), or hormones (steroids hormones). Proteins can be functional (enzymes or hormones) and involved in metabolism or structural, where they form part of a cell or organism's structure. Nucleic acids govern cell activity through protein synthesis [see Unit E].

5. Monosaccharides, disaccharides, and polysaccharides are carbohydrates that differ in terms of the number of single sugar molecule they contain. A monosaccharide is simply a single sugar molecule, like glucose or fructose; a disaccharide is a combination of two single sugars, like maltose, and sucrose; and a polysaccharide contains many sugars, like cellulose, starch and glycogen.

6. The empirical formula of carbohydrates is CH_2O. Maltose is $C_{12}H_{22}O_{11}$. It is almost a multiple of the empirical formula, except that a molecule of water (H_2O) is given off as a byproduct when it forms.

7.
Basis of Contrast	Cellulose	Starch	Glycogen
Structure	linear	helical; some branching	branching
Function	forms cell walls	glucose storage	glucose storage
Location	plant cell walls	plant tissues	animal tissues (liver)

8. Amino acids differ from each other in terms of the composition of the "R" group. There are 20 variations of this component ranging from a simple hydrogen atom to complex, even cyclic structures. The rest of every amino acid is the same, with a "N-C-C" backbone. An amino acid is illustrated to the right. In this case, the R group is a hydrogen atom, producing glycine, the smallest amino acid.

9.
Level of Structure	Shape	Bonding
Primary	linear	peptide bonds between consecutive amino acids
Secondary	α-helix; β-pleated sheets	H-bonds between non-consecutive amino acids
Tertiary	3-dimensional ("glob" shaped)	some H-bonding; also a possibility of ionic and/or covalent bonding
Quaternary	varies; involves the association of tertiary structures together to form a complex	weak electrostatic bonding

10. RNA is similar to DNA in that both are polymers of nucleotides. Each of the nucleotides is composed of a base, a sugar and a phosphate group. Their nucleotides, however, are different. The sugar in RNA is ribose; in DNA it is deoxyribose. The bases in RNA are Adenine, Guanine, Cytosine, and Uracil. Uracil is not found in DNA; Thymine exists instead. Additionally, DNA is considered to be a nuclear molecule and is double stranded where RNA is a cytoplasmic molecule that is produced in the nucleus. RNA is single stranded.

11. ATP is an energy storage molecule. The energy is stored as bond energy in the bonds between consecutive phosphate groups. Breaking one of these bonds releases energy for cell use. ATP is illustrated to the right.

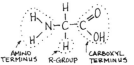

12. a. Ribose is the sugar that is in RNA; deoxyribose is the sugar that is in DNA.
     b. Saturated means there are more hydrogen atoms and no double bonds; unsaturated has less hydrogen and at least one double bond.
     c. The carboxyl group consists of CO_2H; the amine group consists of NH_2.

13. a. Glycine is an amino acid (R = H), and therefore it is a component of proteins.
     b. Glycerol is a 3-carbon alcohol that is a component of neutral fats and phospholipids. Phospholipids are a major component of membranes.
     c. Glucose is a 6-carbon sugar. It is metabolized for energy by mitochondria.
     d. Glycogen is a polysaccharide. It is the storage form of glucose in the liver.
     e. Guanine is a nitrogenous base and is a component of both RNA and DNA. As such it helps make up genetic material.

14. a. A buffer is a molecule that helps a system maintain a constant pH through its ability to utilize or release hydrogen and/or hydroxide ions. An example of a buffer in the human body is bicarbonate ions.
     b. Dissociate means to break apart into ions. Water dissociates into hydrogen and hydroxide ions.
     c. Polypeptides are large molecules that are made out of amino acids. Two different polypeptide sequences combine to make insulin (a protein).
     d. Helix describes the shape of a molecule. Both DNA and proteins are helical-shaped molecules.
     e. Cohesion means "stick together". Water molecules throughout the body are cohesive.

UNIT B – CYTOLOGY

CONCEPT CHECK-UP QUESTIONS

B.1

1. The advantage of compartmentalization in cells is that different reactions and processes can occur simultaneously within the same cell and without interference from each other.
2. Enzymes, made at ribosomes, more into a nucleus to conduct nuclear processes. Messenger RNA and ribosomal subunits are both made in the nucleus, yet function in the cytoplasm.
3. A complete ribosome is a two-part, non-membranous organelle. The parts are made independently by the nucleolus, a region of the nucleus. Each part is a combination of protein and ribosomal RNA. Once out of the nucleus, the parts combine into a complete ribosome in order to function.
4. A polysome is a cluster of the same kind of ribosomes in the cytoplasm that function in association with each other to mass-produce a particular protein.

B.2

1. RER and SER are structurally distinct because RER has ribosomes embedded in its membranes, where SER does not. In terms of function, RER accepts the proteins made at its membranes and prepares them for packaging into transition vesicles. SER contains enzymes (proteins) that allow it to manufacture steroids and phospholipids.
2. An RER connection to the nuclear membrane would allow the movement of ribosomal subunits out of the nucleus and into the membranes of the RER as well as the movement of proteins that are made at the RER into the nucleus. An RER connection to the SER would allow the movement of proteins (enzymes) into the SER either for metabolic processes or for packaging into vesicles.
3. The contents of a transition vesicle (TV) differ from those of a secretory vesicle (SV) due to the function of the Golgi bodies. The Golgi bodies accept the contents of a TV, and prepare them for secretion by concentrating them and sometimes chemically altering them and repackaging them in to an SV.
4. All enzymes are proteins. All proteins are made by ribosomes. The enzymes in a lysosome originate at the ribosomes of the RER (though lysosomes are actually produced by the Golgi bodies). The enzymes are hydrolytic enzymes and are used for breaking down molecules by hydrolysis.

B.3

1. The inner membrane surface area is the location of the some of the chemical reactions of cellular respiration. More surface area means more reactions can take place.
2. The energy for phosphorylation comes from the chemical bond energy released by the hydrolysis of glucose molecules, a process that is completed by mitochondria.
3. The products of cellular respiration (by mitochondria) are the reactants of photosynthesis (by chloroplasts) and vice versa.
4. The cytoskeleton is made out of types of proteins specialized into microtubules and microfilaments.

MULTIPLE CHOICE QUESTIONS:

1. B	5. A	9. A	13. A	17. B	21. B	25. D
2. D	6. C	10. C	14. D	18. C	22. D	26. B
3. B	7. D	11. A	15. A	19. D	23. B	27. A
4. D	8. D	12. B	16. B	20. B	24. C	28. A

WRITTEN ANSWER QUESTIONS:

1. Steroid hormones are produced in the interior of the SER. From here, they follow the secretory pathway. They are put into transport vesicles through the process of blebbing. Once in the vesicle they are moved through the cytoplasm to the Golgi apparatus. The steroids are concentrated and (perhaps) chemically modified into a final product form. The Golgi apparatus produces a new vesicle with the final products. This vesicle is a secretory vesicle. It is moved to the cell membrane where it fuses in such a way as to allow the exocytosis of the steroids.

2.
   a. Lysosomes contain enzymes for hydrolytic reactions in cells. These enzymes (proteins) are manufactured at the ribosomes.
   b. Transport vesicles fuse with the Golgi apparatus so the contents of the vesicles can be modified as required and prepared for secretion. The Golgi apparatus then produces secretory vesicles containing the final products so they can be transported to the cell membrane for secretion.
   c. Secretory vesicles fuse with the cell membrane and exocytosis follows. The cell membrane is also capable of forming vesicles (e.g., food vacuoles) containing materials from the cell's environment (endocytosis).
   d. Ribosomes are the location of protein synthesis. If the proteins being synthesized have to be transported through the cytoplasm (such as those for exocytosis), the ribosomes are embedded in the rough endoplasmic reticulum. The ER makes vesicles for transporting the proteins. The proteins may also be able to move directly from the RER to the SER as with the case of enzymes for steroid synthesis.

3. Intracellular digestion by hydrolytic reactions is the way cells enzymatically digest materials that have been taken into the cell by endocytosis. The process involves food vacuoles, which contain the food materials. These vacuoles are made from the cell membrane. They fuse with lysosomes, which contain the hydrolytic enzymes. The digestive products enter the cytoplasm and are used by the cell.

4. The following are sample answers; other answers are possible.
   a. (Basis of contrast = location) Chloroplasts are found in plant (and some protist) cells, though never in animal cells. Mitochondria can be found in both plant and animal cells.
   b. (Basis of contrast = process) Endocytosis is the process cells use to take in materials that cannot pass through the membrane in any other way. Exocytosis is the process that cells use to release large product materials, like proteins or steroids that the cell has produced.
   c. (Basis of contrast = structure) SES does not have ribosomes associated with it, whereas RER does. The ribosomes embedded in RER are used for the synthesis of specific proteins.
   d. (Basis of contrast = function) A vesicle is a membranous container used for transporting materials in a cell, such as a transition or a secretory vesicle. A lysosome is a membranous container that contains hydrolytic enzymes used for intracellular digestion.

5. Diagram A is most likely SER, which is a set of interconnected membranous channels that extend though sections of the cytoplasm. One function of SER is the detoxification of the cytoplasm.

Diagram B is most likely RER, which looks somewhat similar to SER, but it has ribosomes associated with it. One function of RER is to receive the proteins that are made by the ribosomes and either put them into vesicles or send them to the SER.
Diagram C is most likely a mitochondrion. Mitochondria make ATP through the process of cellular respiration.
Diagram D is most likely a vesicle (though it is hard to tell). Vesicles are membranous containers that cells use to transport materials.
Diagram E is most likely Golgi apparatus. It appears similar to SER, though the saccules are discrete and not interconnected. One function of Golgi apparatus is the final preparation (modification) of cell products to prepare them for secretion.

UNIT C - DNA AND ITS FUNCTIONS

CONCEPT CHECK-UP QUESTIONS:

C.1
1. Protein, specifically a class of proteins called histones.
2. The distance between the strands is held constant by the combinations of pyrimidines bonded to their complementary purines.
3. The DNA in the chromosomes of the daughter cells would not have the correct genetic information.
4. DNA helicase is a hydrolytic enzyme. It breaks H-bonds between the strands of DNA to expose the bases. DNA polymerase catalyzes the rebuilding of DNA by bonding new nucleotides to the exposed bases with H-bonds and then to each other forming the strands.

C.2
1. mRNA is formed through the complementary base pairing of RNA nucleotides onto the DNA template in the nucleus. The role of mRNA is to take the genetic information from the nucleus (DNA) into the cytoplasm where proteins are synthesized.
2. AGAUACGAG. It stands for arginine – tyrosine – glutamate.
3. Methionine is the "start amino acid". The sequence of bases that result in its incorporation into the protein (initiation) marks the beginning of the gene.

4. Termination is the final step in the synthesis of a protein. It occurs because the last t-RNA molecule does not have an amino acid attachment, resulting in the end of the sequence.

C.3
1. Gene mutations occurring in transcription and translation only affect one cell and, therefore, are insignificant. A gene mutation occurring in replication will affect every cell that is produced from that original cell – having the potential to affect every cell in a developing embryo. Even these, however, may not be noticeable because 1) not all genes in a cell are expressed (active), 2) the mutation might not result in the incorporation of a different amino acid due to the semi-conservative nature of the genetic code.
2. The normal codons would be: UCC AAG ACU, which codes for ser-lys-thr. Taking the substitution into account, the codons would be: UCC AAA ACU, which still codes for ser-lys-thr. There will be no change to the protein.
3. Recombinant DNA is DNA that is manipulated so that it contains genetic material from a different species.
4. Production of medicines, drugs etc. (faster and cheaper), and increasing the amount and variety of the world's food supplies.

MULTIPLE CHOICE QUESTIONS:

1. D	6. B	11. D	16. B	21. C	26. B	31. C	36. D	41. B
2. A	7. A	12. A	17. C	22. C	27. B	32. C	37. A	42. C
3. D	8. A	13. C	18. C	23. A	28. A	33. C	38. D	43. A
4. C	9. B	14. B	19. C	24. A	29. C	34. B	39. A	
5. C	10. A	15. B	20. B	25. A	30. A	35. D	40. A	

WRITTEN ANSWER QUESTIONS:

1. DNA is a double-stranded polymer of nucleotides. Its strands are held together by hydrogen bonding between complementary bases. DNA is distinct from RNA in terms of nucleotide structure (deoxyribose instead of ribose; thymine instead of uracil). DNA is double-stranded where RNA is a single-stranded molecule. Finally, DNA is much longer than RNA (RNA is produced from a single DNA gene).

2. The three functions of DNA and their biological significances are:
   a. Replicate (make copies of itself). This is significant for cell division as each daughter cell requires the correct number of chromosomes.
   b. Control protein synthesis. The specific base sequence of DNA ultimately determines which amino acids will be incorporated into the protein to be formed.
   c. Mutate. The base sequence of DNA can be altered through mutations, which changes the ability of the cell to produce a specific protein, thus impacting some aspect of cell metabolism.

3.

Bases of Contrast	Code	Codon	Anticodon
Location	DNA	mRNA	tRNA
Structure	sequence of many bases (A, T, G, C) comprising a gene	sequence of a selection of three bases (A, U, G, C) complementary to a sequence contained in the code	sequence of a selection of three bases (A, U, G, C) complementary to a sequence contained in the codon
Function	ultimately determines which amino acids should be incorporated into an amino acid that is to be constructed	is the transcribed message for a particular amino acid; takes this message to the ribosomes	transports amino acids to ribosomes during translation

4. The three types of RNA and their functions are:
   a. mRNA – transports the transcribed DNA base sequence to the ribosome for translation.
   b. tRNA – transports amino acids to the ribosome during translation.
   c. rRNA – the component of ribosomes that ensures correct alignment of the mRNA for translation.

5. A start codon marks the beginning of the mRNA strand and thus governs the incorporation of the first amino acid (methionine) into the sequence. Stop codons mark the end of the mRNA strand. The tRNA molecules with the corresponding anticodon for a stop codon do not carry an amino acid. As a result, the assembly of amino acids into the sequence is halted.

6. The DNA segment contains 120 nucleotides. There are 26 Guanine nucleotides, and therefore also 26 Cytosine nucleotides (DNA is double-stranded and these bond together.) As well, there are 34 Adenine nucleotides and 34 Thymine nucleotides, which all adds up to 120.

7. a. Base deletion occurs when an intended nucleotide is left out of the sequence that is being formed. The new DNA strand that is produced is, therefore, one nucleotide short. This has a disastrous effect on the protein that would be coded for by this strand because the triplet codons that would be formed from it would (most likely) be incorrect from the point of the deletion on to the end of the strand.
   b. Base addition occurs when an additional nucleotide is incorporated into the sequence that is being formed. The new DNA strand that is produced is, therefore, one nucleotide longer than it should be. This has a disastrous effect on the protein that would be coded for by this strand because the triplet codons that would be formed from it would (most likely) be incorrect from the point of the addition on to the end of the strand.
   c. Base substitution occurs when the incorrect nucleotide is incorporated into the strand that is being produced. This very often means that the mRNA strand produced from this segment of DNA has a single codon that codes for an incorrect amino acid. If it does, the resulting protein will have an error in it and the molecule will not function as it is intended. However, the Genetic Code is degenerative, meaning that some substitutions, particularly when they are in the third position, do not have any bearing on the protein that is made. In these cases, the protein will be correct and the mutation will not be manifested phenotypically.

8. a. and b. (at right)
   c. Replication is the process where replicas of DNA molecules are made. The resulting molecules are identical because each one is formed from one of the parent strands by complementary base pairing.

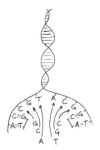

9. a, b, and c.

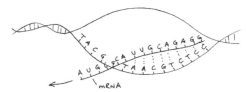

   d. The product of transcription is mRNA. The process occurs in the nucleus, and the newly constructed mRNA leaves through a nuclear pore and travels to a ribosome where it is involved in translation.

10.

   c. The product of translation is protein. The proteins can be considered in two groups depending on the location of the ribosome involved. If the ribosome is free in the cytoplasm or part of a polysome, then the protein will be used in the cytoplasm. If the ribosome is embedded in RER, then the protein will be put into a vesicle and may eventually be secreted like a protein hormone or digestive enzyme (via the secretory pathway) or be put into a lysosome as a hydrolytic enzyme and used for intracellular digestion.

11. Mutations have their greatest impact if they occur during replication when they have the potential to repeatedly produce incorrect proteins (i.e., every time they are used). If incorrect pairing occurs during translation or transcription, then only one molecule of protein is produced incorrectly, and the next time that gene is used, the process more than likely will occur properly.

UNIT D – MEMBRANE STRUCTURE AND FUNCTION

CONCEPT CHECK-UP QUESTIONS:

D.1
1. Membranes can contain proteins, carbohydrates and lipids. Proteins have many functions, such as providing transportation routes for substances across the membrane. Carbohydrates are components of animal cell membranes where they form part of glycolipids and glycoproteins, which function for cell recognition. Lipids, such as phospholipids form the fundamental structure of the membrane providing it with its polar outer surfaces. Cholesterol is another type of lipid that exists embedded in all membranes. It contributes to the flexibility of the membrane.
2. Cells mix freely with water because of the polar nature of the cell membrane. The phospholipid heads have dipoles as does water.
3. "Selectively permeable" implies a greater degree of participation than "semi-permeable". For example, a colander is semi-permeable because water will pass through it where pasta will not. Size and liquid nature are the criteria for permeability in this example. With membranes, other criteria also exist, such as electrical nature and so

on, and passage through the membrane is governed by the presence of specific proteins.
4. No. Some things have complete permeability though membranes (like O_2, CO_2, NH_3 etc.) and pass through simply by diffusion, almost as if the membrane were not even there. The specific permeability of a cell beyond this depends on the presence of proteins (genetic features). Because different cells express different genetic features, their membrane proteins and, therefore their selective permeabilities, can differ.

D.2
1. Diffusion and osmosis are similar because they are both passive processes (no energy is required) as well as they both satisfy the Law of Diffusion. They are different because diffusion involves the movement of substances other than water (solute particles), where osmosis is the movement of the water, itself (solvent). Another

critical difference is that diffusion does not require a membrane like osmosis does.

2. Membranes have selective permeability and, unless the correct proteins for the passage of the substance in question exist in the membrane, it cannot cross. In these situations, osmosis often occurs to balance the concentrations.

3. If a concentration difference doesn't exist, the need for osmosis is eliminated and there will be no net movement of water. If a concentration difference does exist, opposing pressures (such as gravity or electrical potential) may restrict osmosis.

4. Plants wilt because they dehydrate (low turgidity). Watering a plant promotes the hydration of the cells, which increases their turgor pressure making the plant "perkier".

D.3
1. With facilitated transport particles are moving according to the Law of Diffusion (i.e., down a concentration gradient). This is the opposite in active transport where particles are moving against a concentration gradient. Also, facilitated transport does not require energy (from ATP) where active transport does.

2. No. Normally, cells are not freely permeable to Na^{1+}. It normally requires energy and the presence of proteins in order to pass through a membrane. This is an example of active transport.

3. Vesicles, such as food vesicles are made from cell membrane phospholipids, which should decrease the surface area of the cell. The rate at which the vesicles are produced, however, is balanced by the rate at which new phospholipids are added to the membrane through processes such as exocytosis.

4. Doubling the surface area of a (spherical) cell will produce an 8-fold increase in the volume (square function vs. cubic function). The ratio of a cell's surface area to its volume is critical. Cell metabolism occurs in the volume of the cytoplasm, yet the reactants for the metabolic processes enter and products leave a cell through its membrane. Most cells of multicellular organisms can't change size to preserve the SA/V ratio. Unicellular organisms, however, may have the luxury of adjusting their shape in order to avoid altering the SA/V ratio.

MULTIPLE CHOICE QUESTIONS:

1. A	5. C	9. D	13. A	17. D	21. B	25. A	29. D	33. C
2. C	6. C	10. D	14. C	18. C	22. B	26. B	30. C	
3. C	7. B	11. A	15. D	19. A	23. D	27. A	31. D	
4. B	8. A	12. C	16. A	20. D	24. B	28. B	32. A	

WRITTEN ANSWER QUESTIONS:

1. a. Osmosis. Water is a small molecule that passes freely across most membranes according to concentration imbalances.
   b. Facilitated transport. Amino acids are of such a chemical nature and size that they do not pass freely through membranes. Instead, they cross with the assistance of carrier proteins. Cells use the amino acids for protein synthesis, so their concentration in cells is less than in the extracellular fluids, therefore their movement is passive.
   c. Proteins are so large that they require the interaction of vesicles with cell membranes to get out of cells. Examples of this type of movement include: hormone secretion, enzyme secretion, and neurotransmitter release. Cells generally do not take in proteins. They take in the amino acids and make their own proteins.

2. There are several kinds of proteins that can be found embedded in the phospholipid bilayer of a cell. Three examples of them are: receptor sites, which receive environmental molecules causing a response in the cell; enzymes, which react with environmental molecules and integral (or transmembrane) proteins, which move substances through the membrane.

3. (If one assumes that there is no membrane in the solution) their concentration, the temperature, ionic charge and whether or not the solution was being agitated or stirred in any way would affect the rate of diffusion of the ions.

4. a. Deplasmolysis has occurred. Water moved by osmosis into the bag due to the higher concentration of glucose in the bag than outside of it.
   b. The bag is not permeable to glucose. The bag is permeable to water. If either of these statements were not true, the membrane bag would not have swollen to a spherical shape.
   c. According to the Kinetic Molecular Theory the movement of particles is affected by temperature. If the experiment were conducted at 5°C water would move out by osmosis more slowly and it would take longer to reach the same final observation.

5. a. If the concentration of solutes in a cell's environment were to increase, plasmolysis would occur. In this process, water moves out of the cell according to the Laws of Diffusion.
   b. If the concentration of solutes in a cell's environment were to decrease, deplasmolysis would occur. In this process, water moves into the cell according to the Laws of Diffusion. The extent to which this process occurs in countered by the increasing turgidity in the cell.
   c. Both of these changes affect the surface area / volume ratio of the cell. Cells depend on a delicate balance of surface area to volume in order to maintain normal functions.

6. Yes, it does. The result of the movement of particles by diffusion is the evening out of concentration differences. Osmosis has the same effect therefore it is obeying the same natural law.

7. This piece of apparatus is set up so that there is a concentration of starch separated from water by a SPM. Osmosis occurs which dilutes the starch and increases the volume of the starch solution, thus the starch solution goes up the glass tube. The extent to which it rises is a measure of the amount of osmosis. The force of gravity will counteract the force up the tube. Quantifying this force would provide one with osmotic pressure, which is a measure of the force of the water being drawn into the starch solution.

8.

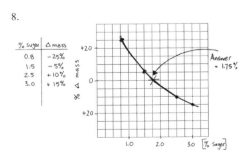

% sugar	Δ mass
0.8	-25%
1.5	-5%
2.5	+10%
3.0	+15%

Answer = 1.75%

UNIT E - ENZYMATICS

CONCEPT CHECK-UP QUESTIONS:

E.1
1. In a general sense, enzymes speed up metabolism. They increase the efficiency of the metabolic reactions in a cell by ensuring that the reactions take place with lower energy requirements and production.
2. Anabolism and catabolism are opposites. Metabolic reactions can either "build up" molecules (synthesis), which is anabolism, or "break down" molecules (hydrolysis), which is catabolism.
3. The shape of an enzyme is of paramount importance to its ability to function. Enzymes have a "sweet spot" called an active site. The conformation (shape) of the active site dictates its ability to combine with substrate molecules and form the activated enzyme-substrate complex, which is required for the enzymatic reaction to take place. If the shape of the active site changes, as it might with temperature or pH shifts, the ability of the substrate to join the active site is affected and the ES complex may not form.
4. Inhibitors get in the way of enzymatic reactions. They prevent the ES complex from forming effectively.

E.2
1. There is probably no real comparison because exothermic metabolic reactions do not occur unless they are conducted with the enzyme. The absence of the enzyme means that the energy requirements to get the reaction started (activation energy) would be too high that it is unlikely that an organism or cell would have this much available energy at any given time. Also, uncatalyzed reactions have the potential of producing large quantities of excess energy (exothermic), which is potentially lethal for cells.
2. Enzymes lower activation energy.
3. Negative feedback is a mechanism where product of a metabolic reaction inhibits the reaction (or the first dedicated step of the reaction) that produces it. In this manner, negative feedback contributes to homeostasis.
4. Co-factors are typically mineral ions (like Ca^{2+}). Their presence in enzymatic reactions has a moderating effect on the movement of

electrons as bonds are broken and formed. (Their larger cousins, the heavy metal ions, disrupt enzymatic reactions due to their large influence on electron distribution). Co-enzymes are very different. They are organic molecules (as opposed to being ions), and have the ability to harbor and release ions, like hydrogen. As such, they supply or accept these ions during the reaction.

E.3:
1. Optimum means "best". The word is correctly used as an adjective to describe a condition that can affect enzymatic reactions. For example, an optimum pH would be the best pH for the reaction to occur at. Below or above that pH there is the probability that the conformation of the active site is affected, therefore the ability of the enzyme to function "optimally" is compromised.
2. The Kinetic Molecular Theory states that at lower temperatures molecules move less. With reduced molecular movement, there are fewer collisions and, therefore, fewer encounters between active sites and substrates. Continued decreases in temperature mean even less molecular movement and, therefore even lower reaction rates.
3. A competitive inhibitor is a molecule that is similar enough in shape to the substrate that it can get lodged in the active site and essentially "plug up" the reaction (i.e., they compete with the substrate for occupancy of the active site).
4. Non-competitive inhibitors and heavy metal ions are very different yet function very similarly. They both disrupt electron distribution in enzymes and therefore affect the ability of the enzyme to catalyze reactions. Non-competitive inhibitors are organic molecules that do this by bonding with the substrate at some location other than the active site. Heavy metal ions may form complexes with the enzymes because of their ionic change. Their large positive nuclei further disrupt electron distribution in the enzyme. In both cases, the enzyme cannot function effectively.

MULTIPLE CHOICE QUESTIONS:

1. A	5. C	9. A	13. A	17. D	21. B	25. C	29. B
2. B	6. C	10. A	14. C	18. C	22. A	26. D	30. C
3. D	7. C	11. C	15. C	19. B	23. C	27. A	31. C
4. C	8. D	12. A	16. B	20. B	24. D	28. D	

WRITTEN ANSWER QUESTIONS:

1. Substrates and inhibitors may be structurally similar, but they are distinct enough that an enzyme will react properly with the substrate, where it will only combine with the competitive inhibitor. If an E-S complex cannot form properly (as in the case of the enzyme combining with the inhibitor), then no reaction can proceed, and no products can be formed.

2. If enzymes are the limiting factor in the reaction, then additional substrates cannot increase the reaction's rate.

3. a. Adding more E_3 will increase the rate that D will be formed (as long as C is not limited and E_4 is not also increased). Also, inhibiting E_4 will mean that more (all?) of Substance C will be converted into D instead of some of it being converted into Substance F.
   b. The substance added must selectively block E_4 from functioning, while not affecting E_3. This substance is likely a competitive inhibitor that blocks the active site of E_4. The inhibitor will not block E_3 from functioning because the active site of E_3 has a different configuration.

4. a. (graph sketched to the right)
   b. There would be approximately 13.5 mL oxygen consumed per hour
   c. Far less oxygen is consumed per hour with Sample A. This can be attributed to the fact that there is no thyroxin in Sample A.
   d. i) 15°C is lower than optimum temperature therefore the reaction rate is lower. The KMT predicts this observation because the molecules move slower and there are fewer collisions between them.
      ii) 35°C appears to be about the optimum, therefore the reaction rate should be the greatest.
      iii) 55°C greatly exceeds the optimum, which suggests that the enzyme should be somewhat denatured allowing for fewer E-S complexes to form, therefore fewer reactions are possible and fewer products can result.

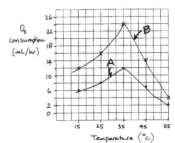

5. a. (graph sketched to the right)
    b. Based on the graph, it should take about 68 seconds to collect the oxygen.
    c. The reaction rate is slowing down from pH 8 to pH 12 as indicated by the increased length of time it takes to collect oxygen, the product. This observation can be explained by the fact that the pH has surpassed the optimum (of about 8.0) and the enzyme catalase is becoming progressively denatured. In this condition, it will not work as well because the active site becomes more distorted and the substrate will not fit properly. Thus the E-S complex will not form as readily, so no products can be formed.

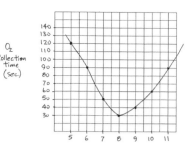

UNIT F – DIGESTIVE SYSTEM

CONCEPT CHECK-UP QUESTIONS:

F.1
1. Ingestion – mouth; digestion – small intestine; absorption – small intestine; defecation – anus
2. a. The endocrine function of the pancreas is producing insulin and glycogen, two hormones that are integral to the body's management of blood sugar levels.
    b. The exocrine function of the pancreas is producing pancreatic juice, which aids in the digestion of food in the duodenum.
3. Firstly, the pancreas produces and releases lipase, the lipid-digesting enzyme. The chemical digestion of lipids is made more efficient when it is already emulsified (physically digested) by bile, a product of the liver. Secondly, the hormones produced by the pancreas affect the storage of sugar by the liver. Insulin tends to increase sugar storage (decrease blood sugar) and glycogen increases blood sugar by causing the release of glucose form the liver.
4. a. digestive functions include bile production; circulatory functions include glucose storage and release, detoxification of blood, removal of worn out red blood cells, production of blood proteins; excretory functions include the conversion of ammonia into urea
    b. These distinctions are arbitrary because of the inter-relatedness of these functions. For example, bile contains by-products of the remains of hemoglobin, which are derived from the destruction of worn out red blood cells and need to be excreted.

F.2
1. a. Physical digestion increases the surface area of ingested food materials in order that chemical digestion can be more efficient.
    b. Chewing (in the mouth), churning (in the stomach), and emulsification by bile (in the duodenum).

2. Salivary amylase digests starch and glycogen into maltose units at a pH of close to 7.0. Pancreatic amylase does the same thing, but at a pH of about 8.4.
3. Pancreatic amylase – starch and glycogen, lipase – lipids, trypsin – polypeptides, and nucleotides – nucleic acids.
4. Protease enzymes are produced and released in the form of precursors (inactive) in order to protect the cells that produce them from their digestive activity. They are activated for use once released. HCl activates pepsinogen in the stomach; trypsinogen is activated enzymatically.
5. Carbohydrates require three chemical digestion steps; lipids require one; proteins require three, and nucleic acids require two.
6. a. Carbohydrates are digested into simple sugars (primarily glucose), lipids into fatty acids and glycerol, proteins into amino acids, and nucleic acids into the components of nucleotides.
    b. No, to consider fatty acids and glycerol as monomers assumes that lipids are correctly called polymers, which they aren't. Also, nucleotides are the monomers of nucleic acids and the digestion of nucleic acids results in the dismantling of nucleotides into their component parts for absorption into the blood.
7. The products of fat digestion enter the lacteals of the villi. In contrast all other products of digestion enter the circulatory capillaries in the villi.
8. Food – bolus – acid chyme – chyme – feces.
9. a. Sphincter muscles control the movement of food materials along the digestive tube, relaxing so that the material may pass through.
    b. Cardiac sphincter between the esophagus and stomach; pyloric sphincter between the stomach and duodenum; ileo-caecal valve between the small and large intestine; and the anal sphincter at the terminal end of the rectum.

MULTIPLE CHOICE QUESTIONS:

1. C	5. C	9. A	13. D	17. D	21. B	25. C	29. B	32. C
2. C	6. D	10. D	14. D	18. B	22. B	26. D	30. A	33. D
3. A	7. A	11. C	15. B	19. B	23. B	27. B	31. D	
4. B	8. B	12. A	16. D	20. C	24. C	28. B	32. D	

WRITTEN ANSWER QUESTIONS:

1. a. Transport: The small intestine has both circular and longitudinal muscles and uses them to conduct peristalsis of food materials along the digestive tract. The small intestine is also connected to the stomach by the pyloric sphincter and to the colon by the ileo-caecal valve, which control the entry and exit of food materials.
    b. Digestion: The small intestine produces enzymes to complete the final chemical digestion of some of the food materials (such as peptidases for protein digestion). The small intestine also receives digestive secretions from other organs (pancreas and gall bladder) to assist with the digestion of food.
    c. Absorption: Both of the ways the small intestine is specialized for absorption are related to surface area. Greater surface areas increase the opportunity for absorption. The small intestine is equipped with villi, which greatly increase the inner surface area, and it is very long increasing the time available for absorption.

2. Bacteria (*E. coli*) in the colon live in a symbiotic relationship with humans. They gain nutrients and have a suitable environment in which to live as well as help us breakdown the food materials, whereby manufacturing vitamins and releasing minerals that we absorb along with water.

3. a. Bile.
    b. Bile breaks lipids apart physically, not chemically like enzymes.
    c. Emulsification increases the surface area of the fat particles (breaks it into smaller fat droplets). Enzymes work at surfaces; the greater the surface, the greater the chemical digestion.

d. The production and release of bile into the duodenum would be ongoing. This would not cause a problem unless the person was on a high fat diet when abnormally large amounts of bile may be required. There is the potential that fat would not be completed digested (enzymes would be inefficient) and would form part of feces.

4. a. The digestion that occurs in the digestive system is extracellular – it occurs in the lumen (interior) of the digestive tract.
   b. The nutrients do not actually enter the body until they have crossed a membrane. This first occurs in the ileum with the absorption at the villi. Until then, the nutrients are located in a tube that runs through the body.

5.

Substance	Source of Substance	Site of Activity	Product of Activity
Ptyalin	Salivary glands	Mouth	Maltose
Pepsin	Stomach	Stomach	Polypeptides
Glucagon	Pancreas	Liver	Glucose
Lipase	Pancreas	Small intestine	Fatty acids and glycerol

6. The first pH change occurs in the stomach as a result of HCl release. Boli usually have a fairly neutral pH when they pass through the cardiac sphincter. The HCl lowers the pH to about 2.5. This low pH kills any bacteria that are in the food and creates a suitable environment for the activity of pepsin (from pepsinogen). The second pH change occurs when acidic chyme passes through the pyloric sphincter and encounters pancreatic juice containing bicarbonate ions. The bicarbonate ions are an excellent buffer in the body and they maintain the pH of the food material (chyme) at about 8.4, which is the optimum pH range for the various food-digesting enzymes that are active in the duodenum.

7. Protein is physically digested in the mouth by chewing and in the stomach by churning. Protein is chemically digested in three steps: The first step occurs in the stomach through the action of pepsin. The products, polypeptide chains, are further digested in the duodenum by trypsin, and are converted into short amino acid sequences called peptides. The peptides are digested in the duodenum by peptidases from the small intestine. The final products are amino acids. Amino acids are absorbed by the villi and they enter the blood stream to be circulated to body cells.

8.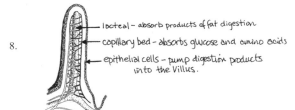

lacteal – absorb products of fat digestion
capillary bed – absorbs glucose and amino acids
epithelial cells – pump digestion products into the villus.

UNIT G – CIRCULATORY SYSTEM

CONCEPT CHECK-UP QUESTIONS:

G.1
1. The pulmonary circuit conducts blood from the right ventricle through the lungs and back to the left atrium; the systemic circuit conducts blood from the left ventricle to all body tissues except the lungs, and back to the right atrium of the heart.
2. The blood returning to the heart from body tissues (systemic) and being pumped to the lungs (pulmonary) is deoxygenated. The blood returning to the heart from the lungs and being pumped into all other body tissues is oxygenated.
3. Valves in the heart (as with valves anywhere in the circulatory system) prevent the backflow of blood.
4. Chordae tendineae are tiny tendons that attach the underside of the AV valves to the muscle mass of the ventricles. Their purpose is to ensure that the AV valves do not invert by being pushed upwards by blood during ventricular contraction.

G.2
1. a. The SA node and the AV node are both involved with the contraction of the heart muscle.
   b. The SA node functions to stimulate the contraction of the atria and to trigger the activity of the AV node. The AV node functions through nerve tissue to cause the simultaneous contraction of the muscle mass of the ventricles.
2. One cardiac cycle includes the contraction of the atria (opening of AV valves; semi-lunar valves are closed) followed by the contraction of the ventricles (during which the atria relax, the AV valves close and the semi-lunar valves open). This is followed by the recovery and relaxation of the ventricles in preparation for the next cycle.
3. The SA node is called the pacemaker because it directly controls the rate of heart contraction. It is connected by the vagus nerve to the medulla oblongata, which can adjust the rate as required.
4. a. High blood pressure can be caused by the retention of too much fluid (usually dietary causes like ingestion of too much salt) or the constriction or obstruction of arteries (caused by many things like stress, smoking, plaque build-up, and so on). High blood

pressure is detrimental because it makes the heart work harder/faster than it should have to.
   b. Low blood pressure is a result of the opposite, namely loss of blood volume (due to injury) or excessive dilation of arteries resulting in less blood getting to essential tissues. This is bad because tissues may starve or be damaged due to the build-up of toxins like ammonia. Low blood pressure may result in lack of kidney function, which requires sufficient blood pressure to remove toxins from the blood (see Unit J).

G.3
1. In the systemic circuit, arteries carry oxygenated blood and veins carry deoxygenated blood, however in the pulmonary system it is the other way around. Blood vessels are categorized (named) by the direction they carry blood rather than by the nature of the blood they conduct. Veins always move blood towards the heart; arteries always move blood away from the heart.
2. There is no pulse in veins because a pulse results from the pumping activity of the heart. The blood pressure in veins is too low and the blood is too far removed from the heart for the heart's activity to have any real impact.
3. Blood is forced back to the heart by body movements (the activity of skeletal muscle). Valves in prevent the backward movement of blood.
4. Sphincter muscles in the circulatory system relax to increase blood flow into a tissue, or contract to reduce blood flow into a tissue. They are located in arterioles.

G.4
1. a. The four anterior arteries branching from the aortic arch are: the left and right subclavians and the carotids. (The right carotid and the right subclavian are united at the aortic arch, but separate immediately.)
   b. The subclavians conduct blood to the shoulders, where they branch to form the brachials, and to the walls of the thoracic cavity. The carotids conduct blood to the head.

2. a. The hepatic portal vein conducts blood to the liver from the intestines. This blood will have variable glucose levels, NH_3 and worn out red blood cells. The liver manages all of these things and returns blood to the vena cava through the hepatic vein. This blood has a constant glucose level, less ammonia (more urea) and fewer worn out red blood cells.
   b. The renal artery can be expected to contain more urea, greater fluid volume and more oxygen (less CO_2) than the renal vein.
3. (Many examples are possible) the brachial (wrist and often the elbow), carotid (neck) and the iliac (groin area)
4. The carotids have nerve receptors that are sensitive to blood pressure and oxygen content.

G.5
1. The umbilical vein conducts oxygenated blood towards the fetus, while the two umbilical veins conduct deoxygenated blood towards the placenta (away from the fetus).
2. The arterial duct conducts blood from the pulmonary trunk to the aortic arch. The oval opening allows blood to flow from the right atrium to the left atrium.
3. The first breath inflates the lungs, which reduces the resistance to blood flow through the pulmonary tissues. This allows a significant volume of blood to fill the left atrium, which forces the oval opening (valve) closed. The blood now travels through the pulmonary circuit. The arterial duct gradually closes over a period of a few days.
4. The fetal circulatory tissues that are no longer used (such as the internal portions of the umbilical blood vessels) atrophy.

G.6
1. a. Red blood cells – transport; white blood cells – combat infection; and platelets – blood clotting
   b. The plasma is the solvent for the particles that are transported as well as the reservoir of water that helps regulate body temperature.

2. The circulatory system removes worn out and otherwise inefficient red blood cells to ensure maximum oxygen transport.
3. a. The presence of foreign antigens causes a certain type of white blood cell to release corresponding antibodies. These antibodies are Y-shaped protein complexes that entrap the antigens (and their host cells) rendering them harmless, and they get destroyed. The purpose of agglutination is to combat infection.
   b. Damage to platelets results in the release of thromboplastin, an enzyme that catalyzes the conversion of prothrombin to thrombin. Thrombin is an enzyme that catalyzes the conversion of fibrinogen to fibrin, an insoluble protein that forms a network with trapped red blood cells (clot). The purpose of clotting is to seal a wound (stop bleeding).
4. The functions of blood are sensitive to pH and temperature changes because they are a factor of protein activity. Denaturing these proteins would affect their function.

G.7
1. Four. It passes into and out of a alveolar cell and into and out of a cell of the capillary wall to get into the plasma.
2. a. Attached to hemoglobin (forming HbO_2)
   b. Oxygen is released because the ability of hemoglobin to hold on to it lessens due to a slightly increased temperature and lower pH in systemic capillaries.
3. At the arteriole end they are under the influence of blood pressure (about 35 or 40 mmHg), and they "leak" through the capillary wall. At the venule end, the blood pressure is greatly lessened but the osmotic pressure is heightened due to the retention of proteins in the plasma. The osmotic pressure draws the materials back in.
4. a. Capillary fluid tissue exchange can contribute to tissue swelling if the capillaries become too leaky and a greater than normal amount of fluid gets forced into tissue spaces at the arteriole end (as happens when histamines are present).
   b. The ongoing activity of the lymphatic system will slowly (eventually) reduce the swelling by drawing the excess fluids into lymph and returning them to the circulatory system.

MULTIPLE CHOICE QUESTIONS:

1. B	7. C	13. C	19. A	25. B	31. A	37. A	43. A
2. B	8. D	14. B	20. C	26. A	32. B	38. A	44. C
3. C	9. D	15. A	21. C	27. C	33. C	39. C	45. D
4. D	10. C	16. B	22. B	28. C	34. D	40. A	46. C
5. D	11. B	17. A	23. D	29. B	35. C	41. C	47. B
6. C	12. A	18. D	24. A	30. C	36. A	42. A	

WRITTEN ANSWER QUESTIONS:

1. a. For blood to get from the fingers to the toes, the blood would have to return to the heart through the brachial vein, subclavian vein and finally the superior vena cava. It would enter the right atrium, then the right ventricle and get pumped through the pulmonary artery to the lungs, where it would get oxygenated. It would return to the heart via the pulmonary vein. It would enter the left atrium, then the left ventricle and get pumped into the aorta where it would be conducted to the iliac artery down the legs to the arteries that serve the toes.
   b. Brain to liver… The blood would leave the brain via the jugular vein and into the superior vena cava. It travels to the right atrium of the heart, then the right ventricle, which pumps it to the lungs. From the lungs, the blood goes to the left atrium through the pulmonary veins. Then it goes into the left ventricle, out the aorta and, finally, into the liver though either the hepatic artery or the hepatic portal vein (after it goes to the intestines).

2. a. Globulins are blood proteins produced by the liver. Examples of globulins are prothrombin and fibrinogen, both required for blood clotting.
   b. Leukocytes are white blood cells, some of which produce antibodies. Antibodies will bond onto foreign cells (identified by their antigens) and cause these cells to agglutinate.

3. The lymphatic system plays a major role in regulating the fluid balance of the body. It will absorb excess fluid that accumulates in tissue spaces and return it to the circulatory system. It also stores white blood cells and antibodies in specializations called lymph nodes. These play a role in the immune system by identifying and destroying antigenic material. Further, it absorbs fats in the digestive system.

4. Systemic cells are bathed by extracellular fluid. The composition of the fluid is constantly being modified by cellular metabolism and capillary fluid exchange. Oxygen diffuses from this fluid into cells. Carbon dioxide and ammonia, which are produced, diffuse into it.

5. Systole involves the simultaneous contraction of the ventricles. During this time the atria are relaxed. The period of time when the ventricles are not contracting is called diastole. During this time, the ventricles recover from their previous contraction and the atria also contract to direct blood into the ventricles in preparation for the next systole.

6. a. The blood at position 1 is oxygenated and contains nutrients as well as low levels of ammonia. In contrast, at position 2, the blood is deoxygenated, contains fewer nutrients and has higher levels of ammonia.
   b. Fluid movement is occurring at position 2 and position 3. This movement differs in two ways. Firstly, the fluid is moving in opposite directions. At position 2, it is moving out of the capillary into tissue spaces. At position 3, it is moving from tissue spaces into the capillary. Secondarily, the movement is a result of blood pressure at position 2 and a result of osmotic pressure at position 3.

7. a. If the foramen ovale did not close completely, then blood entering the right atrium would still be able to pass over to the left side of the heart. This blood is deoxygenated and would be pumped on to the systemic circuit. The baby would lack adequate oxygen delivery to its organs.
   b. If the ductus arteriosus did not seal off completely the blood being pumped to the lungs would be able to enter the systemic circuit. It would result in low blood pressure in the lung tissue and the circulation of partially deoxygenated blood to the organs of the body.

8. Arterioles are equipped with circular (sphincter) muscles that give them the ability to constrict and dilate. By constricting, they can reduce the blood flow to a particular tissue (like to the skin on a cold day in order to conserve heat). Oppositely, they can dilate to increase the blood flow to a tissue bed.

UNIT H - RESPIRATORY SYSTEM

CONCEPT CHECK-UP QUESTIONS:

H.1
1. The four parts of respiration are: breathing (inhaling and exhaling), external respiration (gas exchange at the lungs), internal respiration (gas exchange at the alveoli), and cellular respiration (making ATP in cells)
2. The nostrils are specialized in two ways. Firstly, they have hairs, which filter particulate matter out of the air that gets breathed in. Secondly, they harbour white blood cells, which help destroy pathogenic material and fight infection.
3. a. Large air passageways are held open (prevented from collapse) by cartilaginous rings.
   b. The cartilaginous rings of the trachea are C-shaped to accommodate the flexibility needed for peristaltic activity of the esophagus.
4. Thin (thin-walled) wet (kept wet) capillaries (highly vascularized) stretch (stretch receptors signal exhalation) over numerous (lots of them) surfaces (maximize large surface area).

H.2
1. The respiratory centre is the medulla oblongata. It is in the brain stem. It initiates inhalation in response to elevated concentrations of CO_2 and H^{1+} in plasma.
2. a. Impulses from the medulla oblongata cause the simultaneous contraction of the diaphragm and the intercostal muscles. The diaphragm becomes straighter (flattens; moves down) when it contracts. When the intercostal muscles contract, they move the rib cage up and out.
   b. The combined effect of their contraction increases the volume of the thoracic cavity, which lowers the internal air pressure thus air is drawn in through the trachea.

3. The pleural membranes provide a lubricated, frictionless surface for the lungs tissue to slide against the inner wall of the thoracic cavity. They also seal the lungs from the rest of the body to help ensure that the trachea provides the only passageway in or out of the thoracic cavity. In this way, they help prevent lung collapse. Further, they cause the lung tissue to adhere to the body wall, which helps expand the lungs during inhalation.
4. Inhaled air is warmed to body temperature, humidified to about 100% humidity and cleaned of debris before it reaches the alveoli.

H.3
1. Carbonic anhydrase catalyzes the reversible reaction between water and carbon dioxide as they form carbonic acid (T = 38 °C; pH = 7.35). At T = 37 °C and a pH of about 7.38, this reaction reverses.
2. Depending on the conditions, hemoglobin normally transports oxygen (HbO_2; oxyhemoglobin), hydrogen (HHb; reduced hemoglobin), and carbon dioxide ($HbCO_2$; carbaminohemoglobin). It transports oxygen in between the lungs and systemic capillaries where the conditions are typically about 37 °C with a pH of about 7.38. It transports the other two substances the reverse direction where the conditions are typically about 38 °C with a pH of about 7.35. (Hemoglobin can also transport other substances such as carbon monoxide.)
3. The primary buffering action in the circulatory system is provided by the bicarbonate ions, which are produced by the dissociation of carbonic acid. The pH is also prevented from dropping because excess H^{1+} (also from the dissociation reaction) bonds to Hb.
4. Most of the carbon dioxide is transported as bicarbonate ions, some is bonded to hemoglobin ($HbCO_2$), and a little of it is transported as CO_2 in the plasma.

MULTIPLE CHOICE QUESTIONS:

1. B	4. A	7. D	10. C	13. D	16. D	19. D	22. A	25. B
2. D	5. C	8. D	11. C	14. D	17. D	20. C	23. A	
3. B	6. C	9. D	12. B	15. C	18. A	21. A	24. D	

WRITTEN ANSWER QUESTIONS:

1. Hemoglobin's transport abilities are affected by both temperature and pH. These, coupled with the relative availability of O_2 and CO_2 in the plasma dictate what hemoglobin transports. The metabolic activity of body cells (such as muscle tissue) generates a little heat, warming the blood slightly (about one °C). Also, the pH is slightly lowered due to the presence of metabolic acids such as, carbonic acid, which dissociates to release H^{1+}. The impact of these changes and the fact that CO_2 is more plentiful than O_2, in systemic capillaries, mean that Hb releases the O_2 it was carrying and picks up either CO_2 or H^{1+}.

   In the lung capillaries, the situation is reversed. Oxygen is more plentiful than CO_2, the temperature is slightly lowered and the pH is slightly higher (due to the absence of metabolic activity and acids). Under these conditions, Hb releases the CO_2 or H^{1+} it was carrying and readily accepts O_2.

2. Inhalation is an active process triggered by an elevated level of CO_2 and H^{+1} ions in plasma. This is detected by the medulla oblongata. Exhalation is a passive process triggered by the stretch receptors of the alveoli.

3. Inhalation is triggered at the medulla oblongata primarily by an elevated level of CO_2 and H^{1+} in the blood that goes through it. Its response is to generate nerve impulses to the diaphragm and intercostal muscles (between the ribs). Both of these muscles contract separately, but in unison. The contraction of the diaphragm causes it to shorten into a lowered position (flatten out) and the contraction of the rib muscles cause them to shorten, which moves the rib cage up and out. Both of these actions increase the volume of the thoracic cavity. The thoracic cavity is lined by pleural membranes, which "suction" the lung tissue to the surfaces of the expanding thoracic cavity. This expanded volume creates a vacuum effect (negative pressure), causing air to enter the respiratory passageway. The two layers of pleural membranes are kept moist and they facilitate the movement of the lung tissue against the inner surface of the rib cage without abrasion as the lungs expand. Note that inhalation is "active" (uses ATP).

Exhalation is triggered by the expanded alveoli, which are equipped with nerve receptors that are sensitive to stretch. During the expansion of the lungs, a stretch threshold is reached, which generates an impulse to the medulla oblongata preventing it from continuing its stimulation of the diaphragm and intercostal muscles. No longer stimulated, these muscles relax and return to their elongated positions (domed diaphragm and lowered rib cage). This action puts pressure on the lung tissue, which slides over the inner surfaces of the thoracic cavity as they deflate causing air to move out of the respiratory passageway. Note that exhalation is "passive" (doesn't use ATP).

4. a. In systemic capillaries, carbonic anhydrase catalyzes the reaction between water and carbon dioxide to produce carbonic acid. Carbonic acid readily dissociates into the ions H^{1+} and HCO_3^{1-}. This reaction is summarized as:

$$H_2O + CO_2 \longrightarrow H_2CO_3 \longrightarrow H^{1+} + HCO_3^{1-}$$

   b. In pulmonary capillaries, carbonic anhydrase catalyzes the reverse of the reaction it did in the systemic capillaries, namely the breakdown of carbonic acid, which is reformed from ions, into water and carbon dioxide. This reaction is summarized as:

$$H^{1+} + HCO_3^{1-} \longrightarrow H_2CO_3 \longrightarrow H_2O + CO_2$$

5. Mucus in the air passageways has three functions. The most obvious is its role of catching air borne substances and removing them from the air so that cilia can sweep them out of the passageways. In addition to this, the mucus keeps the air passageways moist, which contributes to their flexibility and, therefore their function – particularly when considering the smaller bronchioles and alveoli. (Exhaled water also contributes to this.) Finally, mucus provides the radiant thermal energy to regulate the temperature of the inhaled air.

6. Alveoli are specialized for their function in the following ways:
    a. numerous to increase the surfaces for gas exchange.
    b. thin-walled to maximize the amount of gas exchange.
    c. internally coated with lipoproteins that reduce the surface tension and prevent collapse.
    d. have stretch receptors to signal the end of inhalation and allow exhalation.
    e. moist, which facilitates gas exchange.
    f. highly vascularized for gas exchange between air and blood.

UNIT I – NERVOUS SYSTEM

CONCEPT CHECK-UP QUESTIONS:

I.1
1. CNS = Central NS, PNS = Peripheral NS, ANS = Autonomic NS and SNS = Somatic NS
2. a. Sorting centre = thalamus; directs impulses to the appropriate part of the cerebrum
    b. Homeostatic centre = hypothalamus; monitors internal body conditions and initiates responses to maintain homeostasis
    c. Reflexive centre = medulla oblongata; monitors specific body functions and responds reflexively to changes
    d. Respiratory centre = medulla oblongata; monitors concentrations of CO_2 and H^{1+} in plasma; controls breathing rate
    e. Thinking centre = cerebrum; associates impulses together in different regions to produce integrated thought
    f. Coordination centre = cerebellum; monitors body orientation in space to provide balance and coordination
3. The corpus callosum allows the two hemispheres to communicate with each other
4. The hypothalamus is closely associated with the circulatory system (monitors blood) and endocrine system (causes hormone release)

I.2
1. Sensory neurons can be distinguished from motor neurons in terms of structure, function and location. Structurally, sensory neurons have long dendrites and short axons where motor neurons have short dendrites and long axons. Functionally, sensory neurons conduct impulses towards the CNS, where motor neurons conduct impulses away from the CNS. In terms of location, sensory neurons enter the CNS through the dorsal root, where motor neurons leave the CNS through the ventral root.

2. A neuron is a nerve cell. A nerve is a bundle of the long processes (parts) of neurons. A mixed nerve is a bundle of motor neuron axons and sensory neuron dendrites.

3. Myelin sheath is a fatty tissue constructed out of discrete fatty cells called Schwann cells. These cells are individually wrapped around the long processes of neurons. The points between the Schwann cells are known as the nodes of Ranvier. Myelin sheath functions to speed up impulse transmission and prevent the ions from one neuron from affecting adjacent neurons (cross-communication).

4. a. Reflex arcs involve impulse transmission along three neurons. A stimulus initiates the process in a sensory dendrite, which takes the impulse to the spinal cord and relays it to a motor neuron via an interneuron. The motor neuron conducts the impulse along its axon to an effector.
    b. The advantage of reflex arcs is speed of response. Because association neurons do not take the impulses to the CNS for interpretation and integration (no thinking involved) the required synapses occur in the spinal cord and responses occur more quickly.

I.3
1. Resting potential is the electrical nature of a neuron membrane at rest. It has a charge, roughly –65 mV on the inside relative to the outside. The charge is due to the unequal distribution of ions.
2. a. The sodium gates open in response to stimulations of the membrane that surpass a threshold. Such stimulation can be physical or chemical.
    b. The sodium gates close when the cytoplasm of the neuron becomes +40 mV relative to the outside.

c. The potassium gates open at the same time as the sodium gates close (i.e., when the cytoplasm of the neuron becomes +40 mV relative to the outside).

d. The potassium gates close when resting potential is achieved again.

3. The role of the Na/K pump is to move sodium ions to the outside and potassium ions to the inside of neurons. This "pumping" action restores and maintains the resting potential. It requires ATP.

4. Impulses are all the same magnitude because they result from ion movement when protein gates change shape. The shape changes are a direct result of electrical differences and not a result of the type or strength of stimulus that caused the impulse in the first place.

I.4

1. a. Presynaptic membranes are equipped with contractile proteins and vesicles of neurotransmitters, which are lacking behind postsynaptic membranes. Postsynaptic membranes are equipped with receptor sites for neurotransmitters, which are lacking on presynaptic membranes.

   b. Impulses cannot be transmitted in both directions across synapses because of the differential structure described in part a. The neurotransmitters are only ever released from the presynaptic side and are only ever received at the postsynaptic side.

2. The influx of Na^{1+} across a postsynaptic membrane would initiate an impulse in the receiving cell. As a result, one could expect that the axoplasm would gain a positive charge, causing the sodium gates to close and the potassium gates to open. The potassium ions would rush out ("downswing") and the electrical nature would return to resting potential. Meanwhile, the chemical activity would cause the impulse to move along the second neuron.

3. a. Enzymes in the synaptic gap destroy neurotransmitters and return the synaptic region to its original condition.

   b. When an impulse reaches the presynaptic membrane, the calcium ions diffuse from the synaptic gap into the axoplasm through calcium gates. These ions combine with the contractile proteins, which are attached to the vesicles of neurotransmitters and cause them to shorten thus promoting exocytosis.

4. Inhibitory and excitatory neurotransmitters have opposite effects. Inhibitory neurotransmitters reduce the potential for depolarization, where excitatory neurotransmitters increase the potential for depolarization.

I.5

1. Acetylcholine and noradrenaline are both neurotransmitters but with different purposes. Acetylcholine is the neurotransmitter for the parasympathetic part of the ANS. As such is promotes vegetative body functions. Noradrenaline is the opposite. It is the neurotransmitter for the sympathetic nervous system and it promotes active body functions.

2. a. Noradrenaline is the neurotransmitter for the sympathetic nervous system and it promotes active body functions, where adrenaline is a hormone produced by the adrenal medulla. Chemically, they are very similar molecules.

   b. Noradrenaline and adrenaline have some similar effects. Noradrenaline increases the activity of certain tissues such as the heart and breathing rate. Adrenaline does the same, but its release into blood has more wide spread impacts as it prepares the body for flight or fight.

3. Sympathetic stimulation of the circulatory system speeds up the heart (active body function) where it will slow down the digestive system. Digestion is a passive (vegetative) activity.

4. The hypothalamus causes both the anterior and posterior pituitary to release hormones. The anterior is stimulated by receiving "releasing hormones", which are carried from the hypothalamus by blood. The hormones that are stored in and released by the posterior pituitary are produced by the hypothalamus. Their release is triggered by a nerve impulse from the hypothalamus.

MULTIPLE CHOICE QUESTIONS:

1. A	6. A	11. D	16. D	21. B	26. D	31. A	36. D	41. C
2. A	7. D	12. B	17. A	22. D	27. C	32. D	37. C	42. C
3. D	8. C	13. D	18. C	23. A	28. C	33. C	38. B	43. B
4. D	9. D	14. D	19. D	24. D	29. B	34. A	39. C	44. D
5. A	10. A	15. D	20. B	25. A	30. A	35. A	40. C	45. C

WRITTEN ANSWER QUESTIONS:

1. (to the right)

2. These are two types of autonomic nerve fibres, the sympathetic and the parasympathetic. They each originate from different regions of the CNS. The sympathetic fibres come from the thoracic and lumbar regions of the spinal cord, where the parasympathetic fibres come from the cranial and sacral regions.

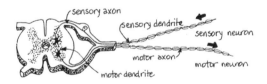

3. a. Region **X** is a synapse. For transmission to occur across a synaptic gap, neurotransmitters have to be released from the axon. The neurotransmitters are often protein-based molecules that are made in the neurons and packaged into secretory vesicles and attached to contractile proteins (cytoskeletal features) by the Golgi bodies. When the impulse arrives at the end of the axon, it affects the presynaptic membrane. Calcium ions (present in the synaptic gap) enter the axon. These ions cause the exocytosis of the neurotransmitters by shortening the contractile proteins and pulling the vesicles to the presynaptic membrane. Once spilled into the gap, the neurotransmitters diffuse through the tiny space. Enzymes destroy many of them. Receptor sites on the postsynaptic membrane take up those that diffuse across the gap. This reception affects the permeability of the postsynaptic membrane and diffusion of ions into the new cell begins. This completes the transmission. The enzymatic destruction of the neurotransmitters continues and the synaptic region regains its original condition ready to be used again.

   b. Region **Y** is a section of an axon, called a node of Ranvier, where an action potential occurs. Disturbance of the membrane in this region of the axon causes a shift in its permeability. This disturbance is a result of the movement of ions as saltatory transmission affects the successive nodes. Once disturbed, the membrane suddenly becomes permeable to sodium ions, which are in abundance outside of the axon. These ions flood to the inside through specialized proteins that are known as "sodium gates". Their arrival in the axoplasm causes the axoplasm to gain a more positive charge. This depolarization shuts the sodium gates and opens the "potassium gates". These gates are also specialized proteins, but these ones allow the potassium ions to leave the axoplasm. The loss of these ions from the inside causes the repolarization of the neuron's membrane. This electrical disturbance affects the adjacent membrane region resulting in a shift in its permeability, and the process continues. In this way, the impulse moves along the neuron. The sodium and potassium ions are returned to their original positions and concentrations through the activity of Na/K pumps, which remain active even when the membrane is properly polarized. It is said that these pumps maintain the resting potential, keeping the membrane ready for future transmissions.

4. The term "neuroendocrine" implies a partnering between the nervous and endocrine (hormone) systems of the body. This partnering occurs with the hypothalamus and the pituitary glands, which extend below the hypothalamus. The nerve tissue of the hypothalamus is connected to the anterior lobe of the pituitary gland by a portal system. The hypothalamus makes and releases certain RH molecules ("releasing hormones"), which travel to the anterior pituitary through the portal system. They each cause the release of an intended hormone by the anterior pituitary. In contrast, the hypothalamus is in communication with the posterior pituitary via a set of neurons. In this case, the hypothalamus differentially produces hormones according to specific body needs, which travel to the posterior pituitary for secretion into the blood.

5. a. Increased stimulation by the sympathetic nervous system results in more noradrenaline release. Noradrenaline accelerates the heart rate and diminishes the activity of the smooth muscles, such as those of the digestive system. In times of stress, the body not only increases its release of noradrenaline, but also the release of adrenaline from the adrenal gland. The effect of adrenaline not only complements that of noradrenaline, but also does more such as preparing the body for danger and emergency situations. This is known as the fight/flight response.
   b. There are several things that the body does in order to respond to emergency situations. These are designed to do things such as maximize blood flow (and therefore nutrient and oxygen delivery) to the skeletal muscles. The depth and rate of breathing is increased (more oxygen available), blood vessels to the skin are constricted and the blood is re-routed to the skeletal muscles, heart rate increases, and so on.

UNIT J – URINARY SYSTEM

CONCEPT CHECK-UP QUESTIONS:

J.1
1. The liver's major excretory role is to remove ammonia (very toxic) from blood and convert it to urea (less toxic), which it returns to blood. The kidneys remove the urea from blood for excretion in urine. The liver plays another role in excretion by ridding the body of the worn out components of hemoglobin molecules, which come from dead red blood cells. The liver adds these molecules (bilirubin and biliverdin) to bile, which means they will be excreted through the digestive system.
2. a. Urine, once it is produced travels through the renal pelvis region of the kidneys, the ureters, urinary bladder, and urethra before it leaves the body.
   b. The renal pelvis is a collection site, which temporarily holds urine until it can be conducted down the ureters. The ureters are the passageway for urine between the kidneys and the urinary bladder. They are equipped with smooth muscles that force the urine along by peristaltic activity. The urinary bladder is a storage organ, which retains the urine until it is excreted. The urethra is the passageway through which urine is excreted.
3. The complex shape of the nephrons (long and meandering) provides a tremendous amount of surface area for the exchange of materials (mostly removal) between the filtrate and the various extracellular fluid environments that nephrons pass through.
4. The renal artery branches into a series of afferent arterioles. Blood from one of these arterioles enters a glomerulus (where filtration occurs) and leaves the glomerulus through an efferent arteriole. The efferent arteriole conducts the blood to the major capillary bed, called the peritubular capillary network. These capillaries surround the nephron structures. Blood from the peritubular capillary network enters renal venules, which conduct it to the renal vein.

J.2
1. Selective reabsorption occurs along the proximal convoluted tubule. In this process, valuable substances in the filtrate that entered as a result of pressure filtration are removed from the filtrate and returned to plasma. These substances include glucose, amino acids, a little water, and some ions (like sodium and chlorine)
2. a. The different processes that occur along the descending and ascending sides of the loop of Henle result from the difference in permeability of these two parts of the tubule. The descending side is water-permeable. As the tubules descend deeper into the medullary region, it encounters an environment that is increasingly hypertonic. Water leaves the filtrate as a result. In sharp contrast, the ascending side is not water-permeable, but is permeable to sodium ions. The filtrate at the bottom (beginning) of the ascending side is rather concentrated due to the loss of water. The surrounding environment is hypotonic to it therefore (on the ascending side) sodium ions (and chlorine ions) leave the filtrate.
   b. The filtrate composition at the top of the two convoluted tubules is actually very similar. The biggest difference is in terms of the amount of urea (much greater at the top of the ascending side).

Along the loop of Henle, the filtrate lost volumes of water as well as salt ions, meaning the total volume of the filtrate is less. Other initial components of the filtrate (glucose and amino acids) were removed before the top of the loop of Henle. The other substances that get added to the urine (like histamines) have not yet been added.
3. a. It is critical for survival that the pH of blood remains relatively constant at about 7.35. This pH is adjusted and regulated at the distal convoluted tubule by the movement of hydrogen and bicarbonate ions between the forming urine and the blood in the peritubular capillary bed. If the pH is low, more H^{1+} will be added to the urine and less HCO_3^{1-} will be absorbed. Alternatively, if the pH of blood is too high, less H^{1+} will be excreted and more HCO_3^{1-} will be recovered from urine.
   b. This question is referring to large molecules like penicillin and histamines. These are pumped out of the blood capillaries and into the distal convoluted tubule region of the nephron (a process called tubular secretion).
4. Urine is made more concentrated by the removal of water along the nephron tubule as it encounters the hypertonic environment in the medulla of the kidney. Urea can be very highly concentrated in the collecting duct as it passes through this region and will diffuse out of the tubule and contribute to the concentration of substances that make up the hypertonic environment (promoting water removal).

J.3
1. Large, frequent urine volumes – due to water retention in the urine resulting from the lack of glucose recovery.
   Thirst – due to dehydration because of the greater volumes of water that get excreted.
   Glucose in urine – because of a lack of ability to recover it from the filtrate.
2. ADH is produced by the hypothalamus and released into blood by the posterior pituitary gland in response to a decreased volume of blood, such as during times of dehydration. This may or may not be associated with low blood pressure as well, depending on the body's activity level at the time. ADH increases the porosity of the collecting duct to water, thus additional water escapes from urine to be recovered by the blood. This increases the concentration of the urine and decreases its volume.
   Aldosterone is produced and secreted into blood by the adrenal cortex in response to low sodium levels in blood (which may also be coupled with low blood pressure). This hormone affects the distal convoluted tubule and increases the retention of sodium (and secretion of potassium). Retaining sodium causes more water to be retained as well. This hormone increases the concentration of urine and decreases the volume of the urine (same effects as ADH, though the mechanism is different).
3. a. Diuretics include coffee and alcohol.
   b. Diuretics increase the volume of urine (and therefore the frequency of urination). Over-consumption of diuretics may lead to dehydration.

MULTIPLE CHOICE QUESTIONS:

1. D	4. A	7. B	10. C	13. B	16. D	19. C	22. A	25. B
2. B	5. B	8. C	11. C	14. B	17. C	20. D	23. A	
3. C	6. C	9. C	12. C	15. B	18. B	21. B	24. D	

WRITTEN ANSWER QUESTIONS:

1. W (adrenal gland) – produces hormones such as adrenaline (from the medulla) and aldosterone (from the cortex).
    X (kidney) – filters wastes (such as urea) from blood and produces urine for excretion.
    Y (ureter) – uses peristalsis to conduct urine to the urinary bladder for excretion.
    Z (urinary bladder) – stores urine until excretion occurs.

2. a. A = renal artery; B = renal vein; C = ureter
    b. A and B both contain blood. The blood in these blood vessels is different in composition in several ways. For example, blood in A is oxygenated, whereas in B it is deoxygenated. Also, the blood in A contains urea, whereas the blood in B contains significantly less urea. The blood also differs in terms of the stability of the pH, and the content of histamines and penicillin and so on.
    c. B contains deoxygenated blood and C contains urine. These fluids are different in their compositions as well. Blood contains blood cells, whereas urine will not normally contain blood cells. The same is true of globulins and other blood components that are too large to pass through the glomerulus. The concentration of urea in the fluid in C is much higher than in B.
    d. at right
    e. at right
    f. The renal pelvis accumulates and temporarily stores the urine that comes down the collecting ducts. This urine has to be transported along the ureter via peristalsis.

3. R = afferent arteriole   U = proximal convoluted tubule   X = distal convoluted tubule
    S = glomerulus   V = loop of Henle   Y = collecting duct
    T = Bowman's capsule   W = peritubular capillary bed

4. Urine is formed by a fairly complex sequence of events that occurs simultaneously in each of the millions of nephrons that are located in each kidney. Structurally, a nephron is a narrow tube that has a cup-shaped end called Bowman's capsule. This structure, situated in the cortical region, encloses a tuft of capillaries called a glomerulus. From Bowman's capsule, the tubule bends and folds (proximal convoluted tubule) before leaving the cortex and going down into the medulla and back up to the cortex. This portion of the nephron is called the loop of Henle. Once back in the cortex, the tubule twists and folds once more (distal convoluted tubule) before joining the collecting duct. Many nephrons can be attached to the same collecting duct. This common duct transports the almost fully formed urine to the renal pelvis.

    It is important to note a few things in order to explain urine formation. Namely, the folding of the tubule in the cortex regions provides more surface area and time for certain processes to occur. Also, the tubule is very thin and has walls that are one cell thick, which facilitates the movement of materials in and out of the nephron in order to make urine. It is also notable that the ascending side of the loop of Henle is specifically designed to pump Na^{+1} out of the tubule and into the medullary region of the kidney, making the salt concentration of this region higher than that of the cortex.

    Urine formation begins at the Bowman's capsule with the filtration of blood through the glomerulus. Blood pressure is high enough that small molecular components of blood are forced through the walls of the glomerulus into Bowman's capsule and, then into the nephron tubule. This fluid is called the filtrate. As the filtrate is pushed (by the ongoing filtration) along the proximal convoluted tubule, the cells of the tubule recover useful components in the filtrate, such as glucose, and pump them back into the blood of the peritubular capillaries. As the filtrate moves down the loop of Henle into the medulla of the kidney, it encounters the high salt concentration, which draws water out of the forming urine and increases its concentration. This water subsequently moves by osmosis into the blood (which was made hypertonic by the filtration process). The permeability of the cells of the nephron changes on the ascending side and no longer is water able to leave the nephron, however, Na^{1+}. Due to its high concentration, it diffuses into the extracellular fluids of the medulla is (Cl^{1-} follows). By the time the filtrate arrives in the distal convoluted tubule it no longer has nutrients in it, still has urea and is much more concentrated than it was before because it has lost both water and salts. The cells of the distal tubule are specialized to make adjustments to the composition of the blood. It is here that the pH of the blood is adjusted and that penicillin and histamines are added to the urine. Once in the collecting duct, the urine again encounters the salty regions of the medulla as it travels to the renal pelvis. This causes even more water to be reabsorbed, concentrating the urine even more.

5. The hypothalamus is sensitive to blood volume and makes the hormone ADH, which is stored in vesicles in the posterior pituitary awaiting secretion. When the hypothalamus registers a lowered volume, it causes the release of ADH from the posterior pituitary gland. The posterior pituitary releases the ADH into the blood. The effect of the ADH is to increase water reabsorption, thereby counteracting the lowered blood volume.

UNIT K – REPRODUCTION UNIT

CONCEPT CHECK-UP QUESTIONS:

K.1

1. Sperm travel through the seminiferous tubules, where they are produced, to the epididymis, along the vas deferens to the urethra, through which they leave the body.
2. a. The brain determines when the puberty should begin. The hypothalamus initiates the process by producing and secreting gonadotropic releasing hormones into specialized capillaries, which conduct them to the anterior pituitary. These hormones influence the anterior pituitary and cause it to release gonadotropic hormones that bring about the changes associated with puberty.
   b. Testosterone promotes the secondary sex characteristics including growth of the primary sexual structures, ancillary hair, larynx and the skeletal frame, as well as development of skin glands and muscles.
3. Sperm originate from the seminiferous tubules; nutrients primarily from the seminal vesicles; fluid from the prostate and Cowper's gland; and buffers from the Cowper's gland.
4. The epididymis stores sperm cells once they are developed and allows them to mature. The acrosome chemically orients the sperm so it swims in the right direction.

K.2

1. a. The ovarian cycle consists of the follicular phase during which a follicle is brought to maturity, ovulation, which is the release of the ovum; and the luteal phase. During the luteal phase, the tissues of the follicle, which remain in the ovary following ovulation, become established as a temporary endocrine gland and produce progesterone and estrogen.
   b. A Graafian follicle is a mature follicle that bulges out from the surface of the ovary ready for ovulation.
   c. The corpus luteum, through its release of hormones, ensures that the uterus is prepared for implantation and the successful beginning of a pregnancy.
2. a. Ova travel from the ovaries, where they are produced, through the oviducts, the uterus, cervix and vagina.

b. The fluids of the female reproductive system transport the ova. They are drawn into the oviducts by the fimbriae, moved along the oviduct to the uterus by the cilia, and out of the uterus to leave the body along with mucus discharge.

3. a. The uterine cycle has a menstrual phase during which blood tissue is released, a proliferative phase, where the endometrium is built up and a secretory phase, where the endometrium is mature and secretes mucus.
   b. The uterine cycle is under hormonal influence. The lack of estrogen and progesterone causes menstruation. The proliferative phase is a largely a result of increased estrogen release and the secretory phase is largely a result of heightened levels of progesterone.
4. The release of gonadotropic hormones from the anterior pituitary is under negative feedback control. Increased levels of estrogen and progesterone diminish the release of gonadotropic releasing hormones form the anterior pituitary. An even higher level of estrogen has a positive feedback relationship with LH causing a sudden "spike" in its release, which results in ovulation.

K.3

1. Implantation followed by the production of human chorionic gonadotropic mark the beginning of pregnancy.
2. The placenta arises from the interface of embryonic tissues and the endometrium. Primarily it is a complex set of capillaries and arterioles and a venule. The capillaries are adjacent to a corresponding set of maternal capillaries. This association is the structure through which gas exchange and nutrient delivery to the embryo occurs. The placenta has no purpose after a baby is born. It is released from the uterine wall and discharged afterwards.
3. The corpus luteum normally degenerates after about 10 days. In the event of a pregnancy, the new hormone called HCG prevents it from degenerating until the placenta is mature enough to produce the required levels of hormones.
4. Uterine contractions cause labour pains. The contractions result from the release of the hormone oxytocin, which occurs due to pressure on the cervix.

MULTIPLE CHOICE QUESTIONS:

1. A	5. C	9. C	13. C	17. D	21. D	25. C	29. D	33. A
2. B	6. D	10. C	14. A	18. B	22. B	26. B	30. B	34. D
3. D	7. D	11. D	15. C	19. C	23. C	27. D	31. D	
4. B	8. C	12. C	16. B	20. A	24. A	28. B	32. D	

WRITTEN ANSWER QUESTIONS:

1.

Basis of Contrast	Sperm	Ova
Site of Production	Seminiferous tubules in testes	Ovaries
Ability to move	Has a flagellum; can swim	None; moved by external forces
Number produced	Millions	One per month
Size	Smaller	Larger

2.

Structure	Name	Function
T	Fimbriae	Assist with ova entry into oviduct
U	Oviducts	Conduct ova to uterus
V	Ovary	Produce ova and hormones (estrogen and progesterone)
W	Uterus	Site of implantation and embryonic development
X	Endometrium	Thickens with blood supply in preparation for implantation
Y	Cervix	Separates uterus from vagina; signals oxytocin release for uterine contractions

3. Sample A was taken on Day 8; Sample B was taken on Day 13; Sample C was taken on Day 1; and Sample D was taken on Day 23

4. a. U = immature follicle   W = Graafian follicle   Y = corpus luteum
   V = developing follicle   X = ovum   Z = ovary
   b. Ovulation (release of the ovum from the ovary)
   c. Structure Y produces progesterone to maintain the endometrium in the event that fertilization occurs.
   d. Thick and lush, but has not yet become fully secretory.

5.

Structure	Name	Function
A	Epididymis	Sperm storage allowing them time to mature
B	Seminiferous tubules	Spermatogenesis
C	Interstitial cells	Testosterone production

6. a. The hypothalamus in females monitors the blood and produces releasing factors and oxytocin.
    The anterior pituitary is affected by the releasing factors and releases LH and FSH.
    The posterior pituitary releases oxytocin.
    b. Neuroendocrine function in females differs from that of males in that the levels of releasing factors from the hypothalamus are changing cyclically (as are the levels of the LH and FSH), where in males, they are rather constant as testosterone levels are maintained and spermatogenesis is ongoing. Also. Males do not produce and release oxytocin as females do (produced by the hypothalamus and released from the posterior pituitary when the cervix is stimulated).

7. Semen includes sperm (gamete required for sexual reproduction), fructose sugar (to nourish the sperm), water (medium for sperm to swim in, and buffers (to maintain a pH suitable for sperm survival).

8.

Hormone	Source	Target	Effect
oxytocin	posterior pituitary	uterus	promotes contractions of the uterus (labour) leading to childbirth
estrogen	ovary; corpus luteum; placenta	various organs; uterus during follicular phase	female secondary sex characteristics and promotes the thickening of the endometrium
progesterone	ovary; corpus luteum; placenta	uterus (primarily)	maturation of the endometrium, causes it to become secretory; prevents its from degenerating and being discharged
HCG	uterus and placenta	ovary	maintains the corpus luteum
LH	anterior pituitary	gonads	promotes hormone release by the gonads; causes ovulation in females
FSH	anterior pituitary	gonads	gamete development
testosterone	testes	various tissues	male secondary sex characteristics

9. A GnRH is a gonadotropic releasing hormone. There are two common ones. They are produced by the hypothalamus and released into tiny blood vessels that conduct them to the anterior pituitary. Each type causes the release of a specific hormone [LH (also called ICSH in males) or FSH (both males and females] that affects the gonads.

10. a. Both terms refer to the release of gametes. Ejaculation is the release of semen (includes sperm) from the penis. Ovulation is the release of ova from the ovaries. A single ejaculate contains millions of sperm (normally), where an ovary only releases one ovum every 28 days.
    b. Follicle and corpus luteum are, for the most part, the same tissue structure, and both produce and release hormones. The follicle exists before ovulation. After ovulation, the remaining tissue becomes the corpus luteum.

11. The hypothalamus and pituitary glands work together to monitor and release reproductive hormones. The hypothalamus produces releasing factors, which travel to the anterior pituitary and cause the release of hormones such as FSH and LH. It also produces oxytocin, which it releases from the posterior pituitary. This relationship between the brain and hormone-releasing tissues has been termed "neuroendocrine".

GLOSSARY OF TERMINOLOGY

absorption – the "taking in" of substance particles, generally in molecular or ionic form. This is the process by which food nutrients enter the villi of the ileum. Absorption requires transport proteins. In the ileum, the transport mechanism is active transport requiring ATP. See also: "reabsorption".

acetylcholine (Ach) – neurotransmitter released by exocytosis into the synaptic cleft at the end of specific neurons, like those of the parasympathetic division of the autonomic nervous system. Ach stimulates the adjoining cell to "pass on the impulse" before being broken down by the enzyme acetylcholinesterase (AchE).

acid – a substance that promotes a pH of less than 7.0 by releasing H^{1+}. Examples of acids include HCl, (a component of gastric juice), and carboxylic acids such as amino acids, and fatty acids. See also "acid group".

acid chyme – the name given to food material within the stomach. It is a combination of swallowed food (boli) and gastric juice, containing HCl.

acid group – (more correctly carboxylic acid group). This is the functional group of an organic acid consisting of CO_2H. The hydrogen atom tends to dissociate as an ion (H^{1+}), hence making it an acid. Both amino acids and fatty acids have the carboxylic acid group.

acrosome – a gel-like cover of the head of a sperm cell. It functions to orient the sperm to swimming towards the ovum and contains enzymes concerned with penetrating the ovum during fertilization.

actin – a type of protein filament that contributes to the contracting mechanisms in muscle and other cell types.

action potential – the measurable aspect of a nerve impulse (as opposed to resting potential). The action potential is a localized effect caused by the sudden movement of Na^{1+} into a nerve cell coupled with the movement of K^{1+} to the outside. See also: "impulse", "resting potential".

activated E-S complex – according to the lock and key analogy and the induced fit hypothesis of enzyme activity, the substrate joins with the enzyme to form and activated substrate-enzyme complex, which facilitates the reaction. Failure of this complex to form properly prevents the reaction from occurring. (See also "lock and key analogy"; "induced fit hypothesis"; "enzyme".

activation energy – the amount of energy that is required by a reaction in order for it to take place. Enzymes lower the activation energy of a reaction.

active site – the portion of an enzyme that combines with the substrate to form the activated enzyme-substrate complex. As the enzyme is a protein, the active site has a specific shape and chemistry, which is determined by the amino acids that border this region. Resultantly, different enzymes have different active sites. An active site is specific to a certain substrate, which has the matching shape and suitable chemistry to make their combination possible.

active transport – a transport mechanism using membrane proteins that requires the input of energy (ATP) as the particles are being moved against the concentration gradient. Examples of active transport include the absorption food materials in the ileum, establishing and maintaining resting potential along neurons, reabsorption of specific substances in the process of urine formation, and the accumulation of iodine by the thyroid gland for the synthesis of thyroxin.

addition – one of three mechanisms of gene mutations. When addition occurs, an additional nucleotide is inserted into DNA during replication. Addition has a disastrous effect on a gene because it displaces the base sequence making the code after the mutation incorrect, which results in flawed mRNA forming during transcription. See also: "deletion", "substitution".

adenine – one of two nitrogenous bases known as purines, which are integral components of all nucleic acids. Adenine is complementary to the pyrimidine base, Thymine (DNA) and Uracil (RNA). See also: "guanine", "purine".

ADH – See "antidiuretic hormone".

adipose – pertaining to fat. Adipose cells, which store neutral fats, make up adipose tissue, a type of connective tissue.

ADP – adenosine diphosphate; forms ATP through the addition of a phosphate group when sufficient energy is available. It is formed by the energy releasing hydrolysis of ATP. See also: "phosphate", "ATP".

adrenal glands – are the endocrine (hormone-producing) glands situated adjacent to the tops of the kidneys in mammals. The adrenal glands have a distinct outer cortical region separate from the inner medullary region. The adrenal cortex produces several hormones including aldosterone, where the adrenal medulla produces epinephrine (adrenaline). See also: "aldosterone", "epinephrine".

adrenaline – See "epinephrine".

afferent arteriole – small artery that delivers blood to the glomerulus, and as such is associated with urine formation. See also: "efferent arteriole", "pressure filtration".

agglutinate – clump together. Antibodies cause the agglutination of recognizable antigens. See also: "antibody", "antigen".

albumin – a blood protein (or globulin) produced by liver cells. Albumin is responsible for establishing the osmotic gradients required for the return of water (and other substances) during capillary-tissue fluid exchange. Albumin, also a major component of egg whites, is often used in experiments involving the hydrolysis of proteins.

aldosterone – a hormone produced by the adrenal cortex under the influence of the anterior pituitary gland. Aldosterone increases the amount of Na^{1+} that is absorbed from urine at the distal convoluted tubule, coupled with an increase in the excretion of K^{1+}. The effect of aldosterone increases water retention from urine, making urine more concentrated and decreasing its volume.

alimentary canal – See "digestive tract"

alkaline – having a pH greater than 7.0. See "base".

all-or-none response – the effect on a neuron by a stimulus that surpasses threshold. The response is termed all or none because it either causes an impulse or it doesn't. All impulses have the same electrical value. See also: "threshold", "impulse".

alpha helix (∞-helix) – one of two possible secondary structure configurations of proteins. Alpha helices are produced when hydrogen bonds form between nearby amino acids pulling them into a coiled fashion. See also "beta-pleated sheet", "secondary structure".

alveoli – microscopic sac-like endings of the bronchioles, forming the lungs. It is through their surfaces that gas exchange for external respiration occurs. They are specialized in several ways to increase the efficiency of gas exchange.

amine group – a functional group of an organic molecule consisting of NH_2. Amino acids have an amine group. See also: "amino acid".

amino acid – the monomer of proteins. All amino acids have the same fundamental structure including an amine group, a carboxyl (or acid) group and a R-group. Yet, there are 20 differing R-groups resulting in 20 different amino acids. Amino acids can be recognized by their "N-C-C backbone".

ammonia – the inorganic molecule, NH_3. Ammonia is produced in cells through the deamination of amino acids. Ammonia passes from cells into extracellular fluid and then into the blood stream by diffusion. Ammonia is toxic in high concentrations. However, the liver converts ammonia into urea, a less toxic molecule. See also "urea".

amphoteric – having the properties of an acid and a base. Buffers, such as bicarbonate ions (HCO_3^{1-}) maintain a steady pH because they have amphoteric properties, i.e., they are able to accept protons to prevent pH from dropping or donate protons to prevent it from rising.

anabolism – the opposite of catabolism; "building up". Metabolic reactions (metabolism) are either anabolic (such as dehydration synthesis reactions) or catabolic (such as hydrolysis reactions). See also: "metabolism", "catabolism".

anaerobic bacteria – bacteria that survive in the absence of oxygen. E. coli (bacteria) present in the colon are anaerobic.

anal sphincter – the circular muscle at the posterior end of the digestive tract, which relaxes to open and constricts to close the anus.

antagonistic – a feature that works in opposition to another feature. The sympathetic and parasympathetic divisions of the autonomic nervous system are antagonistic in their action. See also "autonomic nervous system".

anterior pituitary – an endocrine gland situated below the hypothalamus; partnered with the posterior pituitary. The anterior pituitary gland is under the influence of the hypothalamus and releases several different hormones, including growth hormone and gonadotropins (luteinizing hormone and follicle stimulating hormone). See also "neuroendocrine control center".

anterior vena cava (AKA superior vena cava) – is the vein that delivers blood from the anterior part of the body (above the heart) into the right atrium. The joining together of other veins such as the jugulars and the subclavians form the anterior vena cava.

antibody – a class of Y-shaped protein that is released by a type of white blood cell in response to the presence of foreign antigens. Antibodies bond onto these antigens, causing agglutination. See also: "agglutinate".

anticodon – a triplet of nucleotides on t-RNA. Each anticodon has a complementary codon on m-RNA. Correct alignment of these triplets is required for protein synthesis. See also: "codon", "translation", "t-RNA".

antidiuretic hormone (ADH) – antidiuretics are substances that reduce urine production by causing more water to be reabsorbed from nephrons and retained by the body. The hormone ADH is made in the hypothalamus and stored and released by the posterior pituitary gland in response to low blood volume (dehydration). By opening water channels in the cells that line the collecting duct of the nephron, it promotes additional water removal from the forming urine as it passes along (down) the collecting ducts to the renal pelvis.

antigen – a substance capable of stimulating the release of antibodies (the immune response). In some cases, antigens are simply chemical substances (toxins). In other cases they are cell markers, such as glycoproteins and glycolipids. The interaction between antibodies and antigens is called agglutination. See also: "antibody", "agglutination".

anus – the posterior opening of the digestive tract, protected by the anal sphincter. See also: "anal sphincter".

aorta – the largest artery of the body conducts oxygenated blood out of the left ventricle. Branches of this artery include the coronary, carotid and subclavian arteries before it loops behind the heart and becomes known as the dorsal aorta. The loop portion, called the aortic arch is specialized with aortic bodies containing chemoreceptors sensitive to the pH, oxygen and carbon dioxide content in blood and baroreceptors sensitive to blood pressure. These receptors stimulate appropriate responses to help maintain homeostasis. See also: "carotid bodies", "receptor".

aortic valve – the semi-lunar valve at the beginning of the aorta. When open, during systole, blood is pumped into the aorta from the left ventricle. During diastole, the valve closes to prevent the backflow of the blood from the aorta into the left ventricle.

appendix – a fingerlike vestigial projection from the caecum (first part of the large intestine). No functions are attributed to it.

arterial – pertains to an artery.

arterial duct (AKA ductus arteriosus) – this fetal artery connects the pulmonary trunk to the aorta thus allowing some blood to bypass the nonfunctional lungs. After birth, the inner lining of the arterial duct begins to proliferate and closes the lumen. The duct is still evident in an adult heart as the arterial ligament, ligamentum arteriosum, which attaches the two major arteries together.

arteriole – a small artery that leads to a capillary bed. Arterioles have the same characteristics as arteries, only to a lesser extent. In addition, they are equipped with sphincter muscles, which can constrict to reduce blood flow into the capillary bed.

artery – a blood vessel that conducts blood away from the heart. All arteries of the body are branches (or sub-branches) of the aorta in the systemic circuit, or the pulmonary trunk in the pulmonary circuit.

association neuron (AKA interneuron) – a nerve cell wholly located within the CNS and conducts impulses within the CNS, either from a sensory neuron to a motor neuron as in a reflexive action, or to association areas of the brain for integration.

ATP – adenosine triphosphate. ATP is a specialized RNA nucleotide. The base adenine plus ribose sugar is known as adenosine. When combined with three phosphate groups it is a valuable energy storage molecule for cells because the phosphate – phosphate bonds are high-energy covalent bonds that are easily broken and reformed as the energy needs in cells changes. Cellular respiration is the metabolic pathway that synthesizes most of a cell's ATP; 38 molecules of ATP are generated from each glucose molecule. See also: "ADP", "cellular respiration".

atrio-ventricular (AV) node – one of two pieces of nodal tissue in the heart. The AV node is under the influence of impulses from the SA node. It generates impulses that travel through a nerve bundle (bundle of His) down the septum to a branching set of nerve fibres, called Purkinje fibers. These innervate the ventricles, thus coordinating their contraction. See also: "SA node", "Purkinje fibers".

atrioventricular (AV) valve – large valves made out of connective tissue that allow blood to pass from the atria to the ventricles of the heart, but not the other way. These valves are closed during systole and opened during diastole The AV valves are equippd with chordae tendineae to prevent them from inverting during systole.

atrium (plural: atria) – the receiving chambers of the heart. The atria pass blood along to the ventricles to be pumped out of the heart. They are separated from the ventricles by AV valves.

atrophy – the diminishing of an organ or tissue due to lack of use. The remnants of the umbilical vessels atrophy in a child after birth.

autodigestion ("self digestion") – Some cells are specialized to auto digest through the rupturing of lysosomes. The hydrolytic contents of the lysosomes cause the degradation of other organelles and large molecules in cells. Often times these molecules are used again in another form by other cells. An example is the autodigestion of a tadpole's tail.

autonomic nervous system (ANS) – a sub-division of the nervous system. The ANS is a strictly an effector system consisting of sets of two motor neurons joined by synapses leading to specific effectors (smooth muscles or glands). These neurons are not under voluntary control, functioning involuntarily (autonomically) instead. The ANS is subdivided into the sympathetic and parasympathetic neurons, which have very significant differences. See also: "sympathetic", "parasympathetic".

axon – the part (process) of a nerve cell that extends from and carries impulses away from the cell body. An axon ends at a synaptic cleft with a knob-like structure. A significant amount of impulse research has been done with axons therefore there are specific terms that relate to the structures of an axon, such as axomembrane and axoplasm, though the knowledge obtained through this research can be applied to all neurons.

basal body – a cytoskeletal feature consisting of $9 + 0$ fibrillar structure made out of the protein tubulin, and believed to give rise to cilia and flagella. See also: "cytoskeleton".

base – has at least two distinct meanings. 1) with reference to pH, a base is a substance that, when in solution, promotes a pH of greater than 7.0 due to the abundance of hydroxide(OH^{1+}) ions. An example is sodium hydroxide (NaOH). 2. with reference to nucleic acids, a base is a nitrogenous substance that combines with a 5-Carbon sugar and a phosphate group to produce a nucleotide. There are five different bases in two groups based on their size. See also: "acid", "purine", and "pyrimidine".

beta-pleated sheet – one of two possible secondary structure configurations of proteins. Beta-pleated sheets are produced when hydrogen bonds form between adjacent chains of amino acids forcing them into a folding pattern. See also "alpha-helix", "secondary structure". aY

bicarbonate ion (HCO_3^{1-}) – Bicarbonate ions are dissociation products of particular salts such as sodium bicarbonate (a component of pancreatic juice) and carbonic acid (formed as a result of carbonic anhydrase activity in blood). In both of these cases, the bicarbonate ions buffer their respective solutions. See also "buffer".

biconcave – refers to shape. Biconcave means dished in on both sides. Red blood cells are biconcave.

bilayer – double layered. Cell membranes have a bilayer of phospholipids. See also: "fluid mosaic membrane model".

bile – a liver secretion that passes through the bile duct into the duodenum. Bile is stored in the gall bladder until its entry into the duodenum is stimulated. Bile is an emulsifier of lipids (fats), meaning it mechanically breaks fats into smaller pieces, often called droplets. Effectively, this increase the surface area of the fats, thus increases the efficiency of the enzyme lipase. Bile also contains bilirubin and biliverdin, products of hemoglobin destruction in the liver, which are excreted through the digestive tract.

biochemical – a chemical substance that is important to living organisms. There are four classes of biochemicals that exist as macromolecules: carbohydrates, lipids, proteins, and nucleic acids. See also: "carbohydrate", "lipid", "protein", and "nucleic acid".

biochemical reaction – a chemical reaction involving molecules form one of the four classes of macromolecules. Typically, these reactions are either hydrolytic (catabolic) or synthesis (anabolic) in nature. See also: "hydrolysis", "dehydration synthesis".

birth canal – See "vagina".

blebbing – the process of producing a vesicle as conducted by ER (as in the case of transition vesicles) and Golgi bodies (as in the case of secretory vesicles). See also: "transition vesicle", "secretory vesicle".

blood – a type of connective tissue consisting of plasma and formed elements. In general, blood has four functions: transport of nutrients and wastes, combating infection, clotting, and regulation of body temperature.

blood clot – an accumulation of the insoluble protein fibrin and trapped blood cells. Fibrin is produces through a cascade of chemical reactions triggered by damage to platelets. Blood clots seal damaged areas of blood vessels allowing them time to repair. Normally, blood clots are dissolved over time by the plasma enzyme plasmin. See also: "thrombocyte".

blood pressure – the force that blood (in the lumen of blood vessels) exerts on blood vessel walls. Blood pressure is variable in arteries due to the pumping action of the heart (produces the "pulse") and diminishes with distance from the heart. Blood pressure is measured in units called mmHg (millimeters of mercury). Blood pressure is normally measured in the brachial artery of the arm, where average reading is considered to be 120/80 (systolic over diastolic). Blood pressure is affected by a variety of factors, such as stress, diet, genetics, exercise, the presence of plaque on arterial walls, etc.

blood velocity – the speed of blood through a blood vessel. Blood velocity is highest and most variable in arteries. It is slowest in capillaries, which maximizes the time for capillary-tissue fluid exchange. Velocity in veins is a result of muscle activity as there is very little blood pressure pushing the blood back to the heart.

blood vessel – a tubular structure that carries blood. Generally, there are considered to be three types of blood vessels: arteries, capillaries and veins. See also: "artery", "capillary", "vein".

bolus – a ball of food prepared by the tongue for swallowing. A bolus carries with it some salivary juice, which lubricates its movement down the esophagus and begins the hydrolysis of starch into maltose molecules.

bonding – the combining together of specific atoms. Significant patterns of bonding include: covalent, ionic, and hydrogen bonds. See each bonding pattern for further information.

Bowman's capsule – the specialized ending of a nephron tubule. Each Bowman's capsule has a porous inner surface that envelops a glomerulus and is the recipient of the fluid material that is forced by blood pressure out of the plasma during pressure filtration. The separated solution is called the filtrate. It gets progressively modified to form urine. See also: "pressure filtration", "glomerulus".

brachial artery/vein – branches of the subclavian arteries and veins that serve the arms. The brachial artery is significant because it is most often used to measure blood pressure.

brain – the major organ of the nervous system. It is a mass of nerve cells and support cells with specialized regions that provide the organism with specific abilities. These regions include the cerebrum, cerebellum, hypothalamus etc. For more information about these regions, see each one. The brain is associated with the rest of the body through 12 pairs of cranial nerves and the spinal cord. It, like the rest of the central nervous system, is protected from surrounding bone by the meninges. See also: "meninges", "central nervous system", "spinal cord".

breathing – the mechanical movements of inhaling and exhaling. See also: "inhaling", "exhaling".

bronchi – the branches of the trachea that conduct air to and from the lungs. Like the trachea, they are protected from collapse by cartilaginous rings. Additionally, their ciliated mucus lining serves to help condition inhaled air by providing moisture, establishing body temperature, and removing debris.

bronchiole – the branches of the bronchi. The bronchioles continue to branch into smaller and smaller passageways finally ending at the air sacs called alveoli. The largest of the bronchioles have cartilaginous walls for support, but, as they get smaller and smaller, they lose their cartilage and rely more on surface tension to prevent collapse during inhalation. There are millions of bronchiole endings and (therefore) millions of alveoli. The entire air passageway (trachea, bronchi, and bronchioles) is called the bronchial tree. See also: "alveoli".

buffer – a chemical substance that resists pH changes by either accepting or donating H^{1+}. Bicarbonate ions (HCO_3^{1-}) are significant buffers in the body. By accepting H^{1+}, a buffer raises the pH; by donating H^{1+}, it lowers the pH. See also: "bicarbonate ion".

bulbourethral gland – See "Cowper's gland".

caecum – the first lobe of the colon (large intestine) immediately following the ileo-caecal valve. The appendix extends from the colon. In some mammals, the caecum is specialized and accommodates cellulose-digesting bacteria. In humans, it is not specialized and does not function any differently than the rest of the colon.

calcium ion (Ca^{2+}) – a significant co-factor (mineral) in the body. Included in its functions are: the role it plays in the conversion of prothrombin to thrombin during the formation of a blood clot; the hardening of bone tissue (calcification); triggering the exocytosis of neurotransmitters during synaptic transmission. See also: blood clotting", "synaptic transmission".

capillary – the smallest type of blood vessel. The walls of capillaries are only one cell thick, thus facilitating capillary-tissue fluid exchange. The collection of capillaries between an arteriole and a venule is known as a capillary bed. Sphincter muscles in the arterioles regulate the amount of blood flowing through a capillary bed. See also: "arteriole", "capillary-tissue fluid exchange".

capillary-tissue fluid exchange – the exchange of fluid materials between blood in a capillary and extracellular fluids in tissues as the blood travels through a capillary bed. At the arteriole side of the capillary bed, the blood pressure is still sufficiently high (35 mmHg) to force water and some small substances out of the capillaries into the tissue spaces. Oxygen and nutrients are readily transported down their respective concentration gradients into the cells. Meanwhile, this movement creates an osmotic gradient that draws water back into the capillary at the venule end. The water that is returned now transports substances that are found in abundance in the extracellular fluids, namely CO_2 and NH_3. Under normal circumstances, there is no net change in fluid volume of the blood.

carbaminohemoglobin – the combination of carbon dioxide and hemoglobin. This is the second most common way that carbon dioxide is transported safely in plasma from systemic tissues to the pulmonary capillaries where it is exhaled. See also: "carbon dioxide", "carbonic anhydrase".

carbohydrate – one of four classes of molecules that form macromolecules. Carbohydrates include sugars (such as glucose, ribose, maltose, etc.) and polymers such as starch, cellulose and glycogen. Carbohydrates have the empirical formula: CH_2O.

carbon dioxide – a toxic byproduct of cellular respiration that is removed from the body by the respiratory system. It diffuses into blood during capillary-tissue fluid exchange and gets transported in a variety of ways in blood: mainly as bicarbonate ions produced by the activity of carbonic anhydrase, secondly bonded to hemoglobin producing carbaminohemoglobin, and a little is transported as a dissolved gas. The medulla oblongata is sensitive to the concentration of CO_2 in plasma and initiates breathing when it surpasses a threshold level. See also: "carbonic anhydrase", "carbaminohemoglobin",and "medulla oblongata".

carbonic anhydrase – an enzyme located in the membrane of red blood cells. The enzyme catalyzes the reversible reaction between H_2O and CO_2 forming carbonic acid. Carbonic acid is not very stable in plasma and readily dissociates forming H^{1+} and bicarbonate ions. The pH of blood is prevented from lowering because the H^{1+} bonds to hemoglobin forming reduced hemoglobin (HHb) and the bicarbonate ions act as a buffer. This forward reaction occurs as part of internal respiration. External respiration (at lungs) involves the reverse reaction, which is promoted by the raised oxygen concentration in pulmonary capillaries, a slightly lowered temperature and raised pH. See also: "external respiration", " internal respiration".

carboxyl group – See "acid group".

carboxylic acid – a hydrocarbon with a carboxyl group. See also "acid group".

carcinogen – a type of mutagen that causes cancer. When a genetic mutation occurs to a gene that regulates a cell's normal division, the cell is no longer able to respond to the signals from surrounding cells. Abnormal, irregular, rapid cell division may result. This is cancer. See also: "mutagen".

cardiac – refers to the heart.

cardiac cycle – the cycle of one heart beat. Generally, (although it is a cycle that has no starting point), the contraction of the atria is considered to be the beginning of the cycle. Their contraction is under the influence of the SA node. It is coupled with the opening of the AV valves because of the forward pressure and blood is moved into the ventricles. Normally, the ventricles are allowed to fill for approximately 0.04 s until the AV node and the Purkinje fibers signal them to contract. Their contraction forces the AV valves shut and the blood is pushed past the semilumar valves into the arteries. This is followed by the recovery of the ventricles and the filling of the atria in preparation for the cycle to repeat itself. The entire cycle takes about 0.85 s (occurring about 72 times per minute, a normal pulse). The SA node is considered to be the pacemaker as it is responsible for the intrinsic nature of the heartbeat. The ANS affects the SA node. See also: "sino-atrial node", "atrio-ventricular node", and "pulse".

cardiac output – the volume of blood pumped out of the heart by each ventricle in one minute. Typically, a ventricle pumps about 70 mL of blood 72 times per minute resulting in a cardiac output of about 5.0 L per minute.

cardiac sphincter – the muscular constriction along the gastro-intestinal tract located where the esophagus meets the stomach (in close proximity to the heart, hence the name cardiac"). This sphincter must relax to allow a bolus to enter the stomach.

carotid artery – a branch of the aorta conducting blood to the head. There is a right and a left carotid artery. They are coupled with the jugular veins that conduct deoxygenated blood away from the head.

carotid bodies – nerve receptors in the carotid arteries that are sensitive to chemical composition and pressure of blood. They trigger impulses to the CNS as required to help maintain homeostasis. See also: "aorta/aortic bodies".

carrier protein – AKA "transport protein". These are transmembrane proteins that are specialized to move substances across membranes. In the case of facilitated transport, the substances are moved down a concentration gradient with out the expenditure of energy. In the case of active transport, the opposite is true.

cartilage – a type of connective tissue that supports structures that need to remain somewhat flexible (external ears for example). It plays a significant role in the respiratory system by supporting and holding open the soft tissues that make up the trachea, bronchi and larger bronchioles. In the tracheal region, the rings of cartilage are known as C-shaped, facilitating swallowing in the esophagus, which lies dorsal to the trachea. The air passageways need this support to prevent their collapse during inhalation. See also: "inhalation", "trachea".

catabolism – the opposite of anabolism, "breaking down". Metabolic reactions (metabolism) are either anabolic (such as dehydration synthesis reactions) or catabolic (such as hydrolysis reactions). See also: "metabolism", "anabolism".

catalyze – to speed up. Enzymes catalyze, or speed up metabolic reactions. See also: "enzyme/enzymatic activity".

cell body – the portion of a neuron where the nucleus is located. The cell bodies of interneurons and motor neurons are located within the CNS, thus producing the "gray matter", while the cell bodies of sensory neurons are located in dorsal root ganglions, outside of the CNS. See also: "dorsal root ganglion" gray matter".

cell membrane (AKA plasma membrane) – the outer membrane of a cell. As with all membranes, the cell membrane had a fluid mosaic structure regulating the movement of materials across it, in this case in and out of the cell. The specific permeability, receptive, enzymatic, and transport abilities of the membrane are a direct result of the actual proteins in the phospholipid bilayer. See also: "fluid-mosaic membrane model".

cell Theory – the modern Cell Theory is based on research that was conducted in the 1800's. It includes two fundamental statements, namely that living organisms are cellular in nature and that cells reproduce.

cell wall – the supportive structure outside of the cell membrane in some cells. For plant cells, deposits of the linear polysaccharide, cellulose, compose the cell wall. For other cell types (bacteria, fungi, and some protists), the composition of the cell wall shows some variation. The hard nature of plant cell walls provides rigidity to plant structures (high turgidity). This support lessens when there is a lack of water and plats wilt. Animal cells lack cell walls. See also: "cellulose", "turgor pressure".

cellular respiration – the metabolic pathways sometimes referred to as carbohydrate metabolism because the initial substrates are carbohydrates, such as glucose. It is through this complex set of reactions that carbon compounds are taken apart and the released bond energies are retained by high-energy phosphate-phosphate bonds of ATP. The reactions require oxygen and produce water and carbon dioxide as waste products that get excreted. Though the initial steps of carbohydrate metabolism occur in the cytoplasm, most of the ATP is generated by subsequent reactions that occur in mitochondria. The products of cellular respiration provide the reactants for photosynthesis. See also: "mitochondria, "ATP", "carbon dioxide".

cellulose – a polysaccharide made from a linear arrangement of glucose molecules. Digesting cellulose requires a particular enzyme to break the bonding pattern linking the glucose molecules together. Humans lack this enzyme and, therefore, cannot digest cellulose. It is also known as dietary fibre, helping to form feces. Cellulose is significant in plant tissues as it forms the supportive cell walls. See also: "cell wall", "dietary fibre".

central fissure – the deep fold between the right and left cerebral hemispheres, which divides the cerebrum in two. The corpus callosum marks the base of the central fissure. See also: "corpus callosum", "cerebrum".

central nervous system (CNS) – the part of the nervous system composed of the brain and spinal cord; distinct from the peripheral nervous system. Sensory information is conducted from the periphery to the spinal cord for reflexive actions and/or to the brain for interpretation. Links (synapses) to other neurons occur within the CNS. See also: "peripheral nervous system", "brain", and "spinal cord".

centriole – a cylindrical shaped cytoskeletal feature linked to the formation of spindle fibres for cell division and to the production of basal bodies. See also: "cytoskeleton", "basal bodies".

centrosome – the region of the cytoplasm close to the nucleus where centrioles are located. See also: "centriole".

cerebellum – the large dorsal lobe of the brain located at the top of the spinal cord. It is not divided into right and left portions like the cerebrum. The cerebellum coordinates body movements and balance.

cerebral – refers to the cerebrum.

cerebrum – the large anterior part of the brain, sometimes referred to as the "thinking part". The central fissure divides the cerebrum into right and left hemispheres. Each hemisphere consists of a set of four lobes, each one having specific functions related to the interpretation and integration of sensory input. Motor responses that result from this integration originate in the cerebrum. See also: "brain", "central fissure".

cervical – refers to "neck". See "cervix".

cervix (AKA neck) – anatomically, there are two regions that are called "cervix": The neck itself, hence the term cervical vertebrae, and the opening to the uterus on the vaginal side. The uterine cervix has a couple of significant functions. Firstly, it secretes mucous, which lubricates and protects the walls of the vagina. Secondly, with reference to childbirth, it is sensitive to pressure and stimulates the secretion of oxytocin, the hormone that promotes contractions leading to childbirth. See also: "uterus", "oxytocin".

channel protein – a transmembrane protein that provides a route (channel) through the phospholipid bilayer of a membrane for the passage of water, thus facilitating osmosis. See also: "osmosis", "fluid-mosaic membrane model".

chemical digestion – the hydrolytic reactions that reduce food molecules to monomers by enzymatic activity. Chemical digestion occurs in the mouth (salivary amylase), stomach, (by pepsin), and the duodenum (by many different enzymes). See also: "duodenum", and the individual enzymes for details or their function.

chemoreceptor – a nerve ending that is sensitive to a specific chemical. There are chemoreceptors located in many places of the body such as in the digestive system that respond to the chemical composition of food materials, in the aortic arch, carotid arteries, and medulla oblongata that respond to various aspects of blood composition. The senses of smell and taste also depend on chemoreceptors. See also: "aortic bodies", "carotid bodies", "medulla oblongata", and "hypothalamus".

chlorophyll – a pigment molecule that is sensitive to light energy. See "chloroplast".

chloroplast – the eukaryotic organelle that contains chlorophyll. The chlorophyll is located in complex membranous structures inside the chloroplasts. Photosynthesis occurs in chloroplasts. See also: "photosynthesis".

cholesterol – a type of steroid that is normally produced by the liver and distributed about the body by the circulatory system. It forms an integral part of cell membranes, providing them with a certain degree of flexibility. Some people's diets contain high amounts of cholesterol, which provides the body with more than it needs and can use. Excess cholesterol remains in the circulatory system and build up on artery walls forming plaques (hard spots), which decreases the flexibility of the arteries. This leads to high blood pressure and increases the chances of a heart attack.

chordae tendineae – small tendons which attach the AV valves to muscular extensions from the inside walls of the ventricles. These tendons prevent the AV valve flaps form inverting during systole (ventricular contraction).

chromatin – See: "chromosome".

chromosome – genetic material consisting of DNA (for most organisms) and small proteins called histones. During cell division, chromosomes become visible because the DNA is wound up on histones, thus creating more massive structures. Normally, however, chromosomes are "unraveled" (called chromatin) and their various genes ensure proper enzyme production as required by the cell. Replication of DNA occurs when it exists as chromatin. See also: "replication".

chromosome mutation – mutation that occurs to the structure of a chromosome, as opposed to changes in nucleotide sequences within genes, which are called gene mutations. Examples of chromosome mutations include the loss of a chromosome or a part of a chromosome during cell division, and gaining of an additional chromosome due to flawed chromosome migration during cell division, which is what happens in cases of trisomy 21 (Down's Syndrome). See also: "gene mutation". XC

churn – muscle activity of the stomach caused by its three distinct layers of smooth muscles (circular, longitudinal, and transverse). The churning action mixes acid chyme and contributes to the physical digestion of food. See also: "stomach", physical digestion".

cilia – short hair-like structures made from 20 molecules of the protein tubulin (9 + 2 fibrillar structure). Basal bodies are believed to produce and anchor cilia as they extend through cell membranes. Cilia line the major air passageways (trachea and bronchi) and are covered by mucous. Like all cilia, these ones move, and in doing so, propel the mucus towards the epiglottis. The mucus traps particulate matter in inhaled air and the cilia move it to be swallowed or otherwise removed from the respiratory system.

clavicle – collarbone; used with reference to the subclavian artery and vein. See also: "subclavian".

clitoris – organ containing erectile tissue associated with the female reproductive system. It becomes engorged with blood during arousal. The clitoris is located ventral to the opening of the urethra.

clone – a genetic copy. With modern technology it is becoming increasingly possible to clone organisms. To date, individuals of many different species have been cloned. Cells taken from individuals can be treated as zygotes, and encouraged to grow and differentiate. At appropriate stages of development the young embryos of mammals are inserted into suitable (surrogate) mothers until birth producing an individual with the identical genetic characteristics of the individual the cell was taken from. Genetic engineering has provided scientists with the capabilities of selecting and combining specific genetic combinations for the zygotes, thus predetermining the characteristics of the offspring. Once mastered, many clones can be made with the same technique. See also: "recombinant DNA".

code – term used to refer to the genetic sequence of nucleotides in DNA. The code determines the sequence of nucleotides produced by replication. See also: "replication", "gene", and "codon".

codon – a set of three nucleotides of mRNA. There are 64 different codons often depicted in a codon chart and referenced to the amino acid that each one is codes for. There are three codons that do not correspond any amino acid, resulting in termination during translation. Codons are complementary to both the segment of DNA that underwent replication resulting in their arrangement and to an anticodon, part of t-RNA that transports the amino acids. See also: "mRNA", "tRNA", "anticodon", and "terminator codon".

co-enzyme – an organic complex, often containing vitamins that assist enzymatic function. Co-enzymes often either accept or contribute ions (atoms) to the reactions as they proceed. Without the availability of the co-enzymes and the particles to be contributed, the reactions will cease. The co-enzyme NAD contains the vitamin niacin and transports (carries) hydrogen for an assortment of enzymatic reactions. See also: "NAD".

cofactor – an inorganic ion that assists enzymatic reactions. Examples include copper and zinc ions. Large metal ions prevent enzymatic reactions from proceeding properly, but these smaller ones help realign and stabilize the electron distribution in the enzymes-substrate complex, thus facilitating the reaction. See also: "enzymatic reaction", "heavy metal ion".

cohesion – means "sticking together". Water molecules are cohesive, meaning they tend to stick together. This characteristic of the nature of water molecules is a result of their polarity and the fact that they form H-bonds with one another. Cohesion should net be confused with "adhesion", which means sticking to something else. See also: "hydrogen bonding".

collagen – a type of protein filament that forms part of connective tissue binding cells together.

collecting duct – the final region of a nephron that urine (from the distal convoluted tubule) flows through to enter the renal pelvis. The level of antidiuretic hormone in the blood affects the collecting duct. More ADH results in increased water reabsorption from the urine, thus concentrating the urine even more.

colon – See "large intestine".

compartmentalization – the subdivision of a larger structure into smaller functional units. For example, a eukaryotic cell has compartments called organelles; multi-cellular organisms have compartments called organs. Compartmentalization increases efficiency by promoting specialization, thus otherwise potentially conflicting reactions and processes can occur simultaneously.

competitive inhibitor – a substrate-like molecule that can occupy an active site and prevent the substrate from forming the activated-enzyme substrate complex with the enzyme.

complementary base pairing – each nitrogenous base, adenine, thymine, guanine, and cytosine, can only pair up with and form hydrogen bonds with one other one due to the location of the dipoles on the molecules. Specifically, in DNA adenine only ever forms two hydrogen bonds with thymine, and guanine only ever forms 3 hydrogen bonds with cytosine. In the case of RNA, the exact same pattern exists, except RNA contains uracil in the place of thymine. As such, each complementary base pair consists of one purine and one pyrimidine.

concentration – a measure of the amount of solute relative to the amount of solvent in a solution.

concentration gradient – the difference in concentration between two regions. Diffusion of a solute occurs from an area of high solute concentration to an area of low solute concentration, i.e., *down* a concentration gradient.

conclusion – a statement of judgment made at the end of a scientific investigation. A conclusion will either support or not support the investigation's hypothesis.

condensation – a change of state from gas to liquid. In biochemistry, condensation reactions involve the removal of water. See also "dehydration synthesis".

conformation – molecular shape. Enzymes, for example, have a precise conformation that determines their activity.

connective – one of four types of tissues (the others being epithelial, nerve, and muscle). Connective tissue includes bone, blood, fat (adipose), etc. Connective tissue is described as the type of tissue that provides support for the body systems. The support may be in physical form (as with bone), maintenance (blood), or storage (fat) etc.

constrict – narrow; used in association with arteries and arterioles. These blood vessels have muscular walls and are able to constrict (decrease the cross-sectional area of their lumen) to reduce blood flow. This causes a simultaneous increase in blood pressure in the constricted vessel. The opposite of constriction is dilation (widening). Arteries and arterioles constrict and dilate as required for specific homeostatic purposes, such as maintaining blood pressure for urine formation (see kidney function), temperature regulation of the blood, redirecting blood (see fight / flight response), and etc.

contractile protein – a cytoskeletal feature comprised of proteins, which display secondary structure. When these proteins associate with specific chemical substances their shape changes. Contractile proteins exist in the terminus (end) of an axon where they attach secretory vesicles to the presynaptic membrane. They are caused to shorten by the inflowing of calcium ions, which ultimately leads to the exocytosis of neurotransmitters.

control – a part of a scientific investigation that differs from the actual experiment by virtue of a single variable. The purpose of a control is to provide data about the absence of the variable thus allowing one to determine the effect of the variable. For example, in an experiment to determine the effect of enzyme concentration on the rate of product formation, varying amounts of an aqueous solution of a particular enzyme would be added to known amounts of substrate. The addition of water to the substrate would be a control. See also: "control group", "controlled experiment".

control group – part of a scientific experiment that uses large sample sizes in order to increase the reliability of the results by decreasing the dependency of the results on a single case, or individual test. For example, in a scientific experiment involving living organisms, the organisms may be divided into two groups, the experimental group and the control group. Both groups are treated the same throughout the experiment with the exception of the variable being tested. The experimental group gets the variable, where the control group does not. See also: "control", "controlled experiment".

controlled experiment – a type of scientific experiment, which includes a control (or a control group). See also: "control", "control group".

copulation – means sexual intercourse. The penis and vagina are the gender-specific organs of copulation.

coronary artery/vein – blood vessels that serve the heart muscle. The coronary artery describes the initial branches of the aorta going directly to the heart muscle as the aorta ascends out of the left ventricle. Similarly the coronary vein is actually a set of veins that conduct blood from the heart tissue to the vena cavae as it enters the right atrium.

corpus callosum – a dense structure lying at the base of the central fissure. As such, it connects the two hemispheres of the cerebrum together. It contains nerve tracts that conduct impulses between the hemispheres facilitating bilateral integration. (corpus means "yellow", callosum means "thick").

corpus luteum – an endocrine structure formed in the ovary out of follicular cells after ovulation, as a result of the surge of luteinizing hormone. The corpus luteum secretes female steroid hormones estrogen and progesterone, but particularly progesterone, which brings the endometrium to maturity by promoting mucus secretions from the glandular tissue found there. Without the presence of human chorionic gonadotropin (HCG), the corpus luteum degenerates after about 10 days. (The natural tissue color of a corpus luteum is yellow; luteum means "body", hence its name "corpus luteum"). See also: "ovulation", "human chorionic gonadotropin".

counter current exchange – an exchange mechanism that occurs when two fluids pass each other traveling in opposite directions. Physiologically, this increases the efficiency of exchanges, which otherwise depend on passive transport mechanisms. This type of mechanism exists between the descending and ascending sides of the loop of Henle, thus adding to the efficiency of urine formation.

covalent bond – a chemical bond formed by the sharing of electrons between atoms (i.e., carbon – carbon bonds are covalent). A single carbon atom will form four covalent bonds, one with each of four other carbon atoms. The covalent bonds between the hydrogen atoms and the oxygen atom in a water molecule are more correctly called polar covalent because of the unequal sharing of the electrons between the atoms.

Cowper's gland – one of three secretory glands associated with the male reproductive system contributing to the production of semen. The Cowper's gland (also known as the bulbourethral gland) is located below the prostate gland and dorsal to the urethra. It produces and secretes a mucous fluid.

cranial nerve – (as opposed to spinal nerve). Humans have 12 pairs of cranial nerves, which connect directly to the brain from the periphery of the body. Some of these nerves are sensory only, such as Cranial Nerve I (olfactory nerve) and II (Optic Nerve), while some are motor nerves, and still others are mixed nerves.

crenation – the process by which cells take on a notched, or toothed appearance due to plasmolysis (the loss of water because of a hypertonic environment). Crenation is most pronounced in red blood cells.

cristae – are the shelf-like folds of the inner membrane of a mitochondrion. The cristae contain enzyme complexes that conduct many of the reactions of cellular respiration. See also: "mitochondrion".

cytology – the study of cells.

cytoplasm – the watery fluid within a cell's plasma membrane in which organelles are suspended. Many reactions and processes occur within the cytoplasm. The amount of water in the cytoplasm is at the mercy of the cell's environment as it gains and loses volume by osmosis.

cytoplasmic streaming (AKA cyclosis) – the phenomenon of the natural fluid movement of the cytoplasm in a cell, particularly plant cells. Cytoplasmic streaming contributes to the visible movement of chloroplasts in algal cells. Because it affects the intracellular

concentrations of substances, it impacts the cells' ability to exchange materials with their environment.

cytosine – one of three nitrogenous bases called pyrimidines. The others are Thymine (T) and Uracil (U). Uracil is found only in RNA and Thymine is found only in DNA, but Cytosine (C) can be found as a component of nucleotides in both DNA and RNA. Cytosine is complementary to Guanine, a purine base.

cytoskeleton – the internal framework of a cell, which consists of protein filaments and strands. Some cytoskeletal features function to maintain cell shape, others allow the shape to alter, and still others serve as scaffolding along which cell organelles move.

defecation (AKA elimination) – the removal of feces from the digestive system.

degenerative – refers to the nature of Genetic Code. A base substitution in the third position of a codon often does not change the amino acid. Because of this, the Genetic Code is said to be "degenerative".

dehydration – removal of water. See also: "dehydration synthesis".

dehydration synthesis – the type of metabolic reaction involving the combining of small molecules, like monomers, together to make (synthesize) a large product. These reactions involve the removal of a molecule of water (dehydration) in order to form the new bond. See also: "hydrolysis".

deletion – one of three mechanisms of gene mutations. When deletion occurs, a nucleotide is omitted from DNA during replication. Deletion has a disastrous effect on a gene because it displaces the base sequence, resulting in "flawed" mRNA forming during transcription. See also: "addition", "substitution".

denature – to change the shape of a protein. The hydrogen bonds that maintain the tertiary structure of a protein are sensitive to environmental conditions such as temperature and pH. When they are broken, the protein can change its conformation (shape). This effect is most significant when enzymes are considered because the slightest shape change can alter the shape of an active site thus preventing the enzyme from functioning.

dendrite – the extension of a neuron that normally conducts impulses towards a cell body. Sensory neurons have long dendrites, where motor neurons have short dendrites.

density – the relationship between mass and volume of a substance. Water has a density of $1.0 \ g/cm^3$. The density of most substances increases with decreasing temperature. The density if water, however is greatest at $4°C$, a property which allows ice to float.

deoxygenated – without oxygen. Blood is deoxygenated in the venous portion of the systemic circuit and the arterial portion of the pulmonary circuit. Deoxygenated blood transports waste products of cellular respiration. See also: " external respiration".

deoxyribonucleic acid – (DNA) the double stranded helical polymer of DNA nucleotides that makes up chromatin (chromosomes). DNA is found in the nuclei of eukaryotic cells. The two strands are held together by H-bonds that form between complementary bases. It is "unraveled" when it undergoes replication and transcription. It winds up (condenses) around small proteins called histones for cell division.

deoxyribose – a pentose (five carbon) sugar ($C_5H_{10}O_4$). DNA contains deoxyribose, which has one less oxygen atom than ribose ($C_5H_{10}O_4$). See also "deoxyribonucleic acid".

dependent variable – in a scientific relationship, the dependent variable is determined by the value of the independent variable. For example, the rate of enzymatic activity depends on temperature. Temperature is the independent variable where the rate of enzymatic activity is the dependent variable. Dependent variables are graphed on the y-axis. See also "independent variable".

deplasmolysis – the gaining of water by a cell when it is subjected to an environment that has a lesser solute concentration than the cytoplasm. This results in increased turgidity in the cell causing it to swell, if possible. This process will cause the hemolysis (bursting) of red blood cells that are placed in distilled water. See also "plasmolysis", "hemolysis", and "turgor pressure".

depolarization – the loss of polarity. Nerve cell membranes are polarized by the action of Na/K pumps, which accumulate Na^{1+} on the outside of the membranes and K^{1+} in the axoplasm. This produces the membrane potential resulting in impulses when the membranes are depolarized. See also "action potential".

detoxify – to remove (destroy) toxins. The smooth endoplasmic reticulum detoxifies the cytoplasm of cells. See also: "smooth endoplasmic reticulum".

diaphragm – the sheet-lie muscle at the base of the thoracic cavity. Contraction of the diaphragm causes it to flatten out from its normal arched shape. This action helps to increase the volume of the thoracic cavity, resulting in inhalation. Exhalation occurs when it relaxes. See also: "inhalation".

diastole – See "diastolic pressure".

diastolic pressure – the pressure that blood exerts outwards on the walls of arteries when the heart is not contracting. Normal diastolic pressure measures about 80 mmHg at the brachial artery. As with systolic pressure, it decreases with increasing distance from the heart. See also: "systolic pressure".

dietary fiber – the human digestive system (as well as others) does not produce the hydrolytic enzyme required to digest the polysaccharide, cellulose. It therefore remains in the lumen of the intestines and helps form feces. For this reason, the term" dietary fiber" is used to designate cellulose. Cellulose is the major component of plant cell walls.

diffuse – See "diffusion".

diffusion – the movement of solute particles in a solvent from an area where they are more concentrated to an area where they are less concentrated, i.e., down a concentration gradient. Diffusion is a passive transport process (does not require ATP). Particles are said to *diffuse*. (verb). See also: "concentration gradient", "passive transport process".

digestive enzyme – hydrolytic enzyme that chemically breaks food molecules into monomers. Digestive enzymes can be found in saliva, gastric juice, pancreatic juice, and intestinal juice.

digestive tract – (AKA "gastrointestinal tract", or "alimentary canal"). This is the tube that conducts food through the body from the mouth to the anus. The tube is specialized into organs and has accessory organs attached, all of which function to digest the food and prepare the undigested material for defecation.

diglyceride – the combination of glycerol and two fatty acids; a neutral fat.

dilate – to widen, as with blood vessels. The relaxation of circular muscles of arteries and arterioles will allow them to dilate, which increases blood flow, and decreases blood pressure.

dipeptide – two amino acids held together by a peptide bond.

dipole – the charge on a polarized molecule; either positive or negative. Unequal sharing of electrons in a polar covalent bond results in a positive dipole (less electron presence) and a negative dipole (greater electron presence). The polarized nature of water due to the unequal sharing of electrons allows it to form hydrogen bonds, which contribute to its unique properties. See also: "negative dipole", "positive dipole".

disaccharidase – the class of enzymes whose substrates are disaccharides. Maltase, sucrase, and lactase are disaccharidases that are found in intestinal juice.

disaccharide – double sugar. The hydrolysis of a disaccharide produces two monosaccharides (single/simple sugars); the dehydration synthesis of two monosaccharides produces a disaccharide. Maltose, lactose, and sucrose are all disaccharides.

dissociate – to break apart into its component ions. The dissociation of HCl produces H^{1+} and Cl^{1-}; H_2O dissociates into H^{1+} and OH^{1-}.

distal convoluted tubule – the part of the nephron between the loop of Henle and the collecting duct. The walls of this portion of the nephron are specialized to regulate the pH and composition of plasma by excreting excess H^{1+}, penicillin, histamines etc. It is also able to absorb specific ions like OH^{1-} and Na^{1+} from the forming urine and is the region of the nephron sensitive to aldosterone. See also: "aldosterone".

diuretic – a substance that increase urine volume. (Coffee is a diuretic.) In contrast, ADH reduces urine volume. See also: "antidiuretic hormone".

DNA – See "deoxyribonucleic acid".

DNA helicase – the enzyme that breaks the H-bonds holding DNA strands together. DNA helicase effectively "unravels" DNA making replication possible.

DNA polymerase – the enzyme that catalyzes the combining of DNA nucleotides into a DNA polymer during replication.

DNA sequence – refers to the sequence of bases (or nucleotides) in DNA. This sequence ultimately determines which amino acids will be incorporated into a protein at a ribosome during translation.

dorsal root – a nerve extension (process) extending out from the dorso-lateral part of the spinal cord. Sensory neurons enter the CNS through dorsal root. Dorsal root ganglia house the cell bodies of sensory neurons.

dorsal root ganglion – the collection of sensory nerve cell bodies located along the dorsal root of a spinal nerve. The ganglion causes a small bulge along the dorsal root.

double helix – describes the twisting spiral shape of DNA. Each backbone strand of DNA is a combination of alternating deoxyribose sugars and phosphate groups. The strands are held together by hydrogen bonds between their bases.

downswing – the portion of the graph of an action potential where the resting potential of the neuromembrane is being restored. The movement of K^{1+} into the cytoplasm when the K^{1+} gates are open causes this effect. See also: "upswing".

ductus (vas) deferens – the tubule connecting the epididymis to the urethra (in males). As such, it conducts sperm to the urethra. The lower portion of the vas deferens, which is still located out of the body wall also stores sperm and promotes their maturation similar to the function of the epididymis.

ductus arteriosus – See: "arterial duct".

ductus venosus – See: "venous duct".

duodenum – the first portion of the small intestine. Food materials (acid chyme, at this stage) are released from the stomach through the pyloric sphincter. The duodenum is the region where the majority of the digestive activity takes place. Bile from the gall bladder enters here, so does pancreatic juice. The duodenal walls are also specialized in that they produce their own set of enzymes (intestinal juice).

E. coli (Escherichia coli) – a common anaerobic bacterium. *E. coli* lives symbiotically in the large intestines of a number of mammals including humans. The presence of *E. coli* is required for the normal production of feces as it begins the decomposition of the fecal material. Through its metabolism, it also produces a small amount of vitamins and releases mineral ions, which are absorbed along with water by the large intestine.

edema – swelling. During normal capillary - tissue fluid exchange, the volume of fluid leaving the capillaries is balance by the volume of fluid returned to the circulator system either directly due to osmotic pressure or indirectly via the lymphatic system. This balance can be upset by the presence of certain chemicals in plasma such as histamines (from basophils, a particular type of white blood cell) Histamines cause an increase in the amount of fluid leaving the capillaries, which is not immediately balanced, resulting in swelling.

effector – refers to either a muscle of a gland. The transmission of an impulse along a motor axon causes an action by an effector.

efferent arteriole – the small artery that conducts blood away the glomerulus, and as such is associated with urine formation. The blood in the efferent arteriole is hypertonic due to pressure filtration. See also: "efferent arteriole", "pressure filtration".

ejaculation – the release of semen.

elastic fibers – components of the walls of arteries and arterioles that give them the property of elasticity, being able to regain their original shape and size after having had a surge of blood pass through them.

elongation – the second of three stages of translation. Translation begins with initiation when the ribosomal subunits come together and starts "reading" the codons. Elongation refers to the sequential addition of amino acids to the growing peptide. The amino acids are transported into position by tRNA molecules. Elongation is followed by termination.

empirical formula – the simplest (reduced)version of a molecular formula. The empirical formula of carbohydrates is CH_2O.

emulsification – the making of an emulsion by the physical suspension of immiscible substances in a solute. Bile is an emulsifier of fats because it breaks them into fat droplets and causes them to be suspended in the aqueous chyme. The advantage of bile as an emulsifier is that it greatly increases the surface area of the fats, thus promoting their chemical digestion by lipase.

emulsifier – a substance that causes emulsification. Bile is an emulsifier. See also: "emulsification".

endocrine – refers to hormones. Hormone-producing glands are called endocrine glands. They are ductless, which means the hormones enter directly into the blood stream and are circulated in this manner to their target organs. See also: "exocrine".

endocytosis – the taking in of substances by a cell involving the breaking of the cell membrane and the subsequent formation of a vesicle in the event of pinocytosis or a fool vacuole in the event of phagocytosis. See also: "pinocytosis", "phagocytosis".

endometrium – the inner lining of the uterus, as opposed to the myometrium, which is the outer muscular portion. The endometrium thickens with blood tissue and becomes glandular as it matures during the uterine cycle. See also: "uterine cycle".

endoplasmic reticulum – an organelle comprised of a network of membranous channels that can be found throughout the cytoplasm. ER, as it is abbreviated, can be divided into two types; smooth and rough, which are distinguished by the absence or presence of ribosomes. See also: "rough ER", "smooth ER".

endothelial cell – the type of cell that lines the inside surfaces of organs. See also: "ductus arteriosus".

endothermic – ("in heat"); some metabolic reactions are endothermic where the products exist at a higher energy state than the reactants. Others are exothermic (give off heat). See also: "exothermic".

environmental mutagen - a feature in the environment that can affect the ability of DNA to replicate properly and therefore allow cells to divide properly. It is known that both radiation and some chemicals are environmental mutagens.

enzymatic activity – the metabolic activity of cells is a result of the action of enzymes. At a more specific level, the activity of an enzyme will constitute one step in a metabolic process. The ability of the enzyme to work properly depends on the conditions like temperature and pH, factors that commonly affect the shape of the active site.

enzyme – a type of protein that has the ability to change another molecule, called a substrate, through its physical association with it. Enzymes have a region called an active site that has a particular conformation (shape) into which the substrate will fit. See also: "activated E-S complex", "active site".

enzyme concentration – one of many factors that can affect the rate of product formation during an enzymatic reaction. The higher the concentration of enzymes, the higher the rate (given that sufficient substrate molecules are present.)

epididymis – the tubular structure along the side of a testis. The seminiferous tubules join together to form the epididymis, and leads into the vas (ductus) deferens. Spermatids mature into sperm during the few days that they spend in these tubules.

epiglottis – the ventral flap of tissue on the top of the trachea (the glottis), just anterior to the larynx. Normally this flap of tissue is opened, however it reflexively closes during swallowing, preventing food materials from entering the air passageways.

epinephrine – a hormone produced by the adrenal medulla and released as a result of sympathetic stimulation during emergency situations. Epinephrine release brings about a series of responses known as the "fight of flight response". See also: "fight or flight response".

epithelial – ("skin"). There are different categories of epithelial cells, based on their shape. A good example of one of these is the columnar epithelial cells that line the villi of the ileum. These cells are specialized to absorb the products of digestion

erectile tissue – tissue that becomes engorged with blood and stiffens as part of the sexual response. Males have erectile tissue in their penis. Females have erectile tissue in their clitoris.

erythrocyte – a red blood cell. Red blood cells originate from bone marrow. When mature, they are biconcave and anucleate. As with other cells, erythrocytes have genetic markers (glycoproteins) called antigens. Differences among glycoproteins constitute the difference between blood types.

esophagus – ("food tube"). The esophagus is located dorsal to (behind) the trachea. It extends from the base of the pharynx to the cardiac sphincter, the entry point to the stomach. The walls of the esophagus are line with both circular and longitudinal muscles that conduct peristalsis to move food material (bolus) along. See also "peristalsis".

estrogen – an ovarian hormone. The production and release of estrogen is under the influence of the gonadotropic hormone, luteinizing hormone (LH). Estrogen has several effects on the female body. It brings about the secondary sex characteristics, promotes breast development and the proliferation of the endometrium (part of the menstrual cycle). At modest levels, estrogen also has a negative feedback relationship with the hypothalamus and the anterior pituitary gland, which releases LH and FSH (gonadotropic hormones). The level of estrogen builds over the first two weeks of the cycle. When its concentration surpasses a threshold,

estrogen is no longer able to depress the release of the gonadotropic hormones. This results in the surge of LH, which promotes ovulation.

eukaryotic ("true kernel") –cells that have a nucleus. In contrast to prokaryotic cells, eukaryotic cells have membranous structures inside of the cell membrane. In addition to a nucleus, they have ER, mitochondria, vesicles, and other organelles.

excitatory neurotransmitter – a chemical substance released by a motor axon, which causes an increase in the activity of the effector. IXY

excretion – the release of waste materials. At a cellular level, excretion occurs as cells lose CO_2 and NH_3 by diffusion. At a large scale, excretion occurs as a result of kidney function, where urine is extracted from blood and stored in the urinary bladder to be released (excreted) when convenient.

exhalation – breathing out, also termed expiration. Exhalation is signaled by stretch receptors on the surface of alveoli. When triggered, they send an impulse to the medulla oblongata, which causes it to stop the impulses to the diaphragm and intercostal muscles of the ribs. When these muscles relax, air is forced out of the lungs (exhaled).

exocrine – a secreting gland whose products are released through a duct (as opposed to endocrine). The enzymes of the digestive system are produced by exocrine glands.

exocytosis – the release of substances from a cell involving the fusion of a secretory vesicle with the cell membrane in such a way that the contents of the vesicle are released to the outside of the cell.

exothermic - ("out heat"); some metabolic reactions are exothermic where the products exist at a lower energy state than the reactants. Others are endothermic ("in heat"). See also: "endothermic").

experiment – a set of procedures designed to test a hypothesis through the controlled manipulation of specific variables.

experimental group – part of an experiment that is not the control group. For example, in a scientific experiment involving living organisms, the organisms may be divided into two groups, the experimental group and the control group. Both groups are treated the same throughout the experiment with the exception of the variable being tested. The experimental group gets the variable, where the control group does not. See also: "control", "controlled experiment".

expiration – See: "exhalation".

external respiration – refers to gas exchange (diffusion) across the alveoli surfaces. External respiration increases the concentration of oxygen and decreases the concentration of carbon dioxide in blood (produces oxygenated blood). See also: "carbonic anhydrase", "internal respiration".

facilitated diffusion – the transport mechanism where particles (substances) move through a membrane according to the Laws of Diffusion and involving membrane proteins. Facilitated diffusion does not require energy. See also: "facilitated transport".

facilitated transport – the movement of particles (substances) across a membrane involving membrane proteins. This movement can be down a concentration gradient (facilitated diffusion) or up a concentration gradient (active transport). See also: "facilitated diffusion", "active transport".

fallopian tubes (oviducts) – the cilia-lined tubular extensions from the uterus leading towards the ovaries ending at the fimbriae. Following ovulation, the ova are directed into the fallopian tubes by the undulations of the fimbriae and then through the fallopian tubes towards the uterus by the cilia – a process that takes a few days. Fertilization of the ova by sperm normally occurs during this time.

false labor – the contractions of the uterus, which precede labor. Labor contractions intensify as a result of oxytocin release, whereas false labor subsides.

fats – a general term for lipids. More correctly, fats refer to the combination of fatty acids and glycerol. Lipids include all hydrophobic molecules, like steroids, waxes, and etc.

fatty acid – a long chain carboxylic acid. Fatty acids can be saturated or unsaturated. See also: "saturated fatty acid, unsaturated fatty acid".

feces – the semi-solid, indigestible material that leaves the anus during defecation.

femoral artery/vein – are the major blood vessels that service the upper leg. The femorals are branches to the iliacs. See also: "iliac artery/vein".

fetal circulation – refers to the circulatory pattern of an unborn mammal. During gestation (development) the lungs do not function for gas exchange. As a result, the fetus must conduct external respiration via the placenta, and have a system that is designed to divert blood away from the lungs. See also: "arterial duct", "oval opening", and "umbilical".

fetus – an unborn mammal.

fibrin – the insoluble protein that snares passing red blood cells to form blood clots. Fibrin is the final product in a cascade of reactions that is initiated by the release of thromboplastin by platelets. See also: "blood clot"

fibrinogen – the precursor of fibrin. Fibrinogen is a globulin (protein produced by the liver) and a normal component of plasma. During the blood clotting process, fibrinogen is converted to fibrin by the action of the enzyme thrombin. See also: "thrombin".

fight or flight response – an autonomic response to danger. This response is characterized by a number of events that better prepares the body to take immediate action (fight or flee). These responses include a redirection of blood from the periphery and the digestive system to the skeletal muscles, increased alertness, heart rate, breathing rate, and etc.

filtrate – the material that passes through a filter. During urine formation, a portion of plasma is forced out of a glomerulus and into Bowman's capsule of a nephron. This fluid called "filtrate" undergoes a number of modifications to become urine. See also "nephron".

fimbriae – the finger-like extensions at the ovarian end of the fallopian tubes. The fimbriae move in an undulating fashion, which encourages ova to enter the oviduct following ovulation.

flaccid (limp) – Erectile tissue (such as in the penis and the clitoris) is flaccid when it is not engorged with blood.

flagella – (plural of flagellum). Flagella are cytoskeletal features made out of the protein tubulin and have the characteristic 9 + 2 doublet arrangement. See: "cilia". Both cilia and flagella are concerned with movement. Each sperm cell has a flagellum, which is often referred to as a tail.

fluid mosaic membrane model – the current model used to describe membrane structure and function. According to this model, membranes are constructed out of a bilayer of phospholipids arranged so that their hydrophobic ends are together. Cholesterol is found in this region as well. Their hydrophilic regions are towards the outsides so they will be miscible with the watery nature of the cytoplasm and the extracellular fluids. This constitutes the "fluid" portion. The mosaic part of the model is because of the seemingly random arrangement and placement of proteins within this bilayer. There are several different types of proteins in membranes including transport proteins, receptor sites, enzymes and etc.

follicle stimulating hormone (FSH) – This gonadotropic hormone is produced and released by the anterior pituitary. Its target organs are the gonads, hence the term "gonadotropic". In males, FSH promotes spermatogenesis, in females it promotes the completion of oogenesis leading to ovulation. The release of FSH is controlled by the hypothalamus through the production of a GnRH. See also: "gonadotropic releasing hormone".

follicles – immature ova. During fetal development, a few thousand cells of the ovaries begin to go through oogenesis. By birth, all the ova a female will ever have are identified. The cell divisions and the production of viable ova are not completed until after puberty and are under hormonal control.

follicular phase – part of the ovarian cycle that is highlighted by the maturation of a single follicle into a Graafian follicle, which bursts under the influence of high levels of LH (luteinizing hormone)

food vacuole – a vacuole (vesicle) formed by endocytosis to encapsulate food particles. In unicellular organisms, food vacuoles will fuse with lysosomes for the hydrolysis (digestion) of the food molecules.

foramen ovale – See: "oval opening".

formed elements – the cellular components of blood that are produced by the body. Formed elements include red blood cells, white blood cells and platelets.

gall bladder – a thin-walled storage sac attached to the underside of the liver. The gall bladder stores bile, which is produced by the liver. When its release is signaled, bile enters the duodenum and emulsifies fats in order to increase the efficiency of lipase.

gamete – a sex cell. Ova (female) and sperm (male) are gametes.

ganglia – a collection of nerve cell bodies outside of the central nervous system. In mammals, ganglia exist along the dorsal roots, next to the spinal cord. Ganglia can also be found in the autonomic nervous system close to the CNS where the synapses between the sympathetic neurons

exist, and in the peripheral tissues where the synapses between the parasympathetic fibers exist.

gastric – pertains to stomach.

gastric juice – the digestive secretion produced by the stomach. Gastric juice is release as a result of the action of the hormone, gastrin. Gastric juice contains HCl, pepsinogen (pepsin precursor), and mucus (mostly water).

gastrin – the hormone produced by the stomach walls. The arrival of food at the stomach triggers the release of gastrin into the blood. Gastrin, circulates the body and allows time for the stomach to churn and physically breakdown its contents before it triggers the release of gastric juice.

gastrointestinal tract – See "digestive tract".

gene – a segment of DNA that codes for a specific protein. A gene undergoes transcription, the first part of protein synthesis, to make mRNA.

gene mutation – a change to the normal nucleotide sequence in a gene brought about by a mutagen (chemical or radiation). Addition, deletion, and substitution are all types of gene mutations. Very often, a mutated gene is unable to code for the correct protein. See also: "addition", "deletion", and "substitution".

genetic disorder – a condition brought on by a genetic mutation.

genetic engineering – the use of technology to make changes to the genetic make-up of an organism. Research in genetic engineering has provided techniques used for making recombinant DNA. See also "recombinant DNA".

globulin – blood protein produced by the liver. Globulins contribute to osmotic pressure during capillary tissue-fluid exchange. Examples of globulins include fibrinogen and prothrombin (blood clotting proteins) and albumin.

glomerulus – a tuft of capillaries lying between the afferent and efferent arterioles in the renal cortex. Fluid is forced from the glomerulus and into Bowman's capsule during urine formation, a process known as pressure filtration. See also: "Bowman's capsule", "pressure filtration".

glucagon – a pancreatic hormone that promotes the breakdown of glycogen into glucose at the liver. Glucagon is produced by the α-cells of the islets of Langerhans. This hormone is released when blood sugar levels are low. Its effect is to increase blood sugar levels. In this way, it opposes the action of insulin.

glucose – the monosaccharide ($C_6H_{12}O_6$) produced by photosynthesis and used for energy (ATP) production in cells through the process of cellular respiration.

glyceride – the combination of glycerol and fatty acid(s). Glycerides can be either mono-, di-, or triglycerides.

glycerol – the three-carbon alcohol that combines with long chain fatty acids to produce neutral fats.

glycogen – a branching polymer of glucose molecules. Glycogen is the polysaccharide that stores glucose in the liver. Its anabolism is promoted by insulin and its catabolism is promoted by glucagons in order to maintain the glucose concentration in the plasma. See also: "insulin", "glucagon".

glycolipid – the combination of an oligosaccharide (short polymer of monosaccharides) attached to a phospholipid on the outer surface of animal cell membranes. These molecular complexes contribute to the cells ability to recognize other cells. See also: "glycoprotein".

glycoprotein – the combination of an oligosaccharide (short polymer of monosaccharides) attached to a protein on the outer surface of animal cell membranes. These molecular complexes contribute to the cells ability to recognize other cells. See also: "glycolipid".

Golgi apparatus (bodies) – cell organelles discovered in the late 1800's by Camillo Golgi (hence it is always capitalized). These organelles appear as a set of flattened saccules oriented such that there is a receiving surface and a releasing surface. They receive vesicles on one side, modify their contents and release newly formed vesicles with modified contents on the other side. These new vesicles are either secretory vesicles containing products to be released or are lysosomes, containing hydrolytic enzymes.

gonad – an organ that produces sex cells. The female gonads are the ovaries and the male gonads are the testes.

gonadotropic releasing hormone (GnRH) – is a hormone (or hormone-like substance) that is produced and released by the hypothalamus. The GnRHs target the anterior pituitary gland and cause them to secrete gonadotropic hormones (LH or FSH).

Graafian follicle – a mature follicle in the ovary prior to ovulation.

gray matter – the darker portion of the CNS that is made up of nerve cell bodies of interneurons and motor neurons in the absence of myelin. The gray matter forms a butterfly-shaped region in the spinal cord when it is viewed in cross-section.

guanine – one of two nitrogenous bases known as purines (adenine is the other one). Guanine is complementary to cytosine (a pyrimidine) and forms three hydrogen bonds with it in a DNA molecule. Guanine is also found in RNA. See also: "purine", "adenine".

HCl – hydrochloric acid. This acid is a component of gastric juice. As such, it converts pepsinogen to the active enzyme, pepsin, which breaks some peptide bonds starting the digestion of proteins in the stomach.

head – one of three main body parts of a sperm cell (along with the tail and midpiece). This is the location of the chromosomes and is covered with the gel-like acrosome. See also: "sperm cell".

heart rate – the number of times a heart contracts in one minute. Left alone, the heart rate is intrinsic, established by the SA node, a specialized tissue, which contracts every 0.85 s or 72 times per minute. The heart rate is influenced by excitatory and inhibitory neurotransmitters. See also: "SA node".

heavy metal ion – ions such as lead and mercury that influence enzyme function because of their large positively charge nuclei. When in close proximity to an enzyme, heavy metal ions can disturb the normal electron arrangement and cause denaturation. See also: "denature".

hemisphere ("half circle") – The central fissure divides the cerebrum into two lateral hemispheres. The corpus callosum interconnects the hemispheres and allows communication between them. See also: "corpus callosum".

hemoglobin – the major transport protein in blood. Hemoglobin is a protein complex that has quaternary structure. The individual polypeptides that make up hemoglobin are associated with an iron-containing unit called the "heme group". Hemoglobin is located in the cell membranes of red blood cells. It is a transport protein, capable of bonding with oxygen, hydrogen, and carbon dioxide, which it does under different conditions. See also: "oxyhemoglobin", "reduced hemoglobin", and "carbaminohemoglobin".

hemolysis – the bursting of red blood cells due to turgor pressure. Turgidity in cells is increased when the cells are in hypotonic conditions. See also: "deplasmolysis".

hepatic – pertains to the liver. Liver cells are called hepatocytes. The hepatic artery conducts oxygenated blood to the liver, the hepatic portal vein takes nutrient-rich blood into the liver from the intestines, the hepatic vein takes blood out of the liver, and the hepatic duct takes bile towards the duodenum.

hepatic portal vein – the major vein that takes nutrient-rich blood from the small intestines to the liver. Portal veins are distinct in that they have capillary beds on both ends. On the intestinal end, this vein receives the glucose and amino acids from digestion and conducts them to the liver. The liver tissue does several things to the blood, including removing excess glucose (above the normal level) and storing it as glycogen. See also: "hepatic vein".

hepatic vein – the major vein that conducts blood from the liver back into the inferior vena cava. On its way through the liver, the blood is treated or conditioned in several ways such as: regulation of the glucose level, removal of damaged blood cells, removal of toxins such as alcohol, and addition of globulins such a blood clotting proteins.

histamine – a protein released as a result of damage to basophils, a type of white blood cell. Histamines promote the movement of fluids from plasma into tissue spaces. As a result, they are responsible for the swelling of tissues. (Antihistamines counteract this effect).

histone – a class of proteins associated with DNA in the make-up of chromosomes. Histones are believed to participate in the regulation of gene activity as well as the coiling up of chromatin to form chromosomes.

homeostasis – the set of process that function to maintain a constant internal environment. See also: "homeostatic mechanism".

homeostatic mechanism – a process that contributes to homeostasis. Homeostatic mechanisms normally work through negative feedback, where a change in the internal environment promotes a counteracting change, thus maintaining and regulating a normal body condition. See also: "negative feedback".

homeostatic regulation – See: "homeostatic mechanism".

human chorionic gonadotropin (HCG) – a small hormone that is produced by the tissues of a placenta at the early stages of a pregnancy. This hormone enters the mother's system and prevents the degeneration of the corpus luteum. This maintains the secretion of progesterone, which is required to prevent menstruation. It is the presence of this hormone that is sought after in a pregnancy test.

hydrogen bond – a weak electrostatic attraction between opposite dipoles, such as those of water molecules. A H-bond forms between the positive dipole of a water molecule (the hydrogen) and the negative dipole of another molecule (the oxygen). A single water molecule can actually form four H-bonds. This makes them cohesive, which contributes to the unique properties of water. Additionally, H-bonds form between successive loops of a protein helping to maintain secondary structure. They also form between opposite dipoles contributing to the tertiary structure of proteins as well as between the complementary bases of nucleic acids.

hydrogen bonding – See: "hydrogen bond".

hydrogen carrier – a molecule that bonds with hydrogen atoms and transports them to a subsequent reaction. For example, NAD, a complex containing the niacin vitamin, gains hydrogen during certain stages of cellular respiration, only to contribute them and their energy to ATP production in subsequent stages. See also: "NAD".

hydrogen ions – protons; when hydrogen ionizes, it loses its electron and gains a positive charge, thus it is abbreviated H^{1+}. The measure of the hydrogen ion concentration can be used to determine the pH of a solution – the more hydrogen ions there are the lower the pH. Hydrogen ions in deoxygenated blood (produced by the activity of the enzyme, carbonic anhydrase) are transported by hemoglobin, thus maintaining a normal pH level in plasma.

hydrolysis – the type of metabolic reaction involving the breaking down of larger molecules (such as a polymers) into their components (monomers) with the addition of water molecules. See also: "dehydration synthesis".

hydrolytic enzyme – an enzyme that conducts a hydrolytic (catabolic) reaction, as oppose to one that conducts an anabolic reaction. Lysosomes contain hydrolytic enzymes for their function of intracellular digestion. See also: "lysosome".

hydrophilic – will mix with water, as opposed to hydrophobic, which will not mix with water. The head of a phospholipid is hydrophilic, yet the fatty acid portion is not. See also: "hydrophobic".

hydrophobic – will not mix with water, as opposed to hydrophilic, which will mix with water. Lipids are defined as molecules that are hydrophobic. See also: "hydrophilic".

hydrostatic pressure – refers to water pressure.

hydroxide ion (OH^{1-}) – The dissociation of water produces equal numbers of hydrogen ions and hydroxide ions. A solution is called a base when the hydroxide ion concentration exceeds the hydrogen ion concentration. Bases have a pH > 7.0. See also: "hydrogen ion".

hypertension – means high blood pressure. See also: "blood pressure".

hypertonic – having a greater concentration than another solution to which it is compared.

hypotension – means low blood pressure. See also: "blood pressure".

hypothalamus – a part of the lower portion of the midbrain. The hypothalamus is responsive to many blood conditions and it produces chemical substances to affect the homeostasis of these conditions. It produces ADH and oxytocin, which get released through the posterior pituitary gland. It also produces GnRHs and other substances that affect the release of hormones from the anterior pituitary. See also: "neuroendocrine control", "anterior pituitary", and "posterior pituitary".

hypothermia – low body temperature (as opposed to hyperthermia). Normal body temperature is considered to by 37°C. This is significant as the majority of enzymes work optimally at or near this temperature. Hypothermia causes a reduction in enzyme function due to decreased molecular activity (kinetic molecular theory), and can lead to death.

hypothesis – a statement that describes a relationship that is believed to be true; an educated guess. A hypothesis gives a starting point for a scientific investigation.

hypotonic – having a lesser concentration than another solution to which it is compared.

ileo-caecal valve – the smooth muscle constriction (sphincter muscle) that controls the movement of chyme from the ileum to caecum (the first part of the large intestine).

ileum – the third and longest part of the small intestine. Villi, for the absorption of food nutrients, line the lumen of the ileum.

iliac artery/vein – the major blood vessels of the legs. The dorsal aorta branches to form the iliac arteries where the iliac veins join together to form the inferior (posterior) vena cava. The umbilical arteries (fetal) are branches of the iliac arteries.

implantation – the embedding of a developing embryo in the wall of the endometrium. After fertilization, an ovum will begin to divide repeatedly by mitosis. By the time it reaches the uterus, it may be a ball of more than 100 cells. These cells are able to secrete enzymes that allow it to embed into the endometrium layer of the uterus. This marks the beginning of a pregnancy. The contact between endometrium and embryonic cells develops into the placenta. These cells secrete the hormone HCG. See also: "placenta", "pregnancy", and "human chorionic gonadotropin".

impulse – an electrical change that moves along a neuron. See also: "action potential".

independent variable – in a scientific relationship, the independent variable determines the value of the dependent variable. For example, the rate of enzymatic activity depends on temperature. Temperature is the independent variable where the rate of enzymatic activity is the dependent variable. Independent variables are graphed on the x-axis. See also: "dependent variable".

induced fit model – a concept that uses the dynamic shapes of molecules to help explain enzyme function. (The same reasoning can be used to help explain facilitated transport). In this model, the union of a substrate into the active site of an enzyme causes bond "stresses" that slightly alter the shapes of the enzyme and the substrate forming the activated enzyme-substrate complex. It is during this stage that the chemical reaction occurs to the substrate. The instability of the complex forces the enzyme to separate from the newly formed products. See also: "activated enzyme-substrate complex".

inguinal canal – the pathway through the embryonic tissues that the testes follow during their movement from inside the body cavity where they are first formed, to the outside where they function.

inhalation – breathing in. When the medulla oblongata registers a significantly high concentration of CO_2 and H^{1+} in blood, it initiates a nerve impulse to the diaphragm and intercostal muscles causing them to contract. Their contraction effectively increases the volume of the thoracic cavity and inhalation follows maintaining a balanced air pressure in the thoracic cavity.

inhibit – prevent or reduce. See also: "inhibitor" in enzyme function, "inhibitory neurotransmitter" in nerve function.

inhibitor – a chemical substance that inhibits an enzymatic reaction. A common type of inhibitor is a competitive inhibitor. See also: "competitive inhibitor".

inhibitory neurotransmitter – a chemical substance that is released by an axon. This substance reduces the activity of the effector that receives it.

initiation (starting) – Translation starts with the formation of a complete ribosome out of the two subunits over the codon AUG, which is recognized as the start codon. AUG codes for methionine, which is the first amino acid in all proteins.

inspiration – See: "inhalation".

insulin – a protein hormone that is produced by the β-cells of the islets of Langerhans in the pancreas. Insulin promotes the uptake and utilization of blood sugar (glucose) by cells. Thus, insulin lowers blood sugar levels. Liver cells convert the glucose into glycogen for storage. Insulin has 51 amino acids, which makes it a fairly short protein. Its effects are counteracted by glucagon. See also: "glucagon".

integral protein (AKA "transmembrane protein") – a protein that is contain within a membrane, often passing from one side of the membrane to the other. Proteins like these participate in the transport of substances from though the membrane by active or facilitated transport.

integration – the combined effect of the transmission of impulses within the CNS by interneurons. The appropriate recognition of stimuli and the simplest actions and thought processes require integration so that they are coordinated and logical.

intercostal (rib) **muscle** – the smooth muscle tissue lying between the ribs. Contraction of these muscles contributes to inhalation in much the same way as the diaphragm. See also: "diaphragm".

internal respiration – gas exchange at the systemic capillary beds. The effect of this gas exchange is to increase the oxygen concentration in the extra-cellular fluids so it can diffuse into the cells. At the same time, CO_2, which diffuses out of cells, enters the blood, making deoxygenated blood. See also: "carbonic anhydrase", external respiration".

interneuron – See: "association neuron".

interphase – the longest stage of the cell cycle. This is when replication occurs. When prophase begins, marking the beginning of mitosis (or meiosis) the chromosomes appear as already duplicated.

interstitial cells – the cells that lie between the seminiferous tubules of the testes. These cells produce testosterone as regulated by luteinizing hormone from the anterior pituitary.

intestinal juice – the digestive secretion of the small intestine. Intestinal juice contains a variety of enzymes that complete the chemical digestion of food. Included in intestinal juice are disaccharidases, peptides, nucleases, and water.

intracellular digestion – digestion within a cell. Food vacuoles, which are formed by endocytosis (phagocytosis), fuse with a lysosome. Lysosomes contain hydrolytic enzymes, which digest the food particles.

intramembranous space – the space enclosed by a membrane.

ion – a charge atom, or group of atoms. Cell membranes are not freely permeable to ions, and require transport proteins to move them across.

ionic bond – a bond between oppositely charged ions. The bond between Na^{1+} and Cl^{1-} in NaCl is an ionic bond. See also: "covalent bond".

islets of Langerhans – clusters of endocrine cells in the pancreas. There are two types of islet cells, α and β. The α-cells produce glucagon and the β-cells produce insulin. See also: "endocrine", "pancreas".

isomers – different arrangements of the same number and kind of atoms forming substances with different chemical properties. Fructose and glucose are isomers, since they are both $C_6H_{12}O_6$.

isotonic – having the same concentration as another solution to which it is compared. See also: "hypertonic", "hypotonic".

jugular veins – the veins that conduct blood from the head down the neck. They join the superior vena cava allowing the blood to enter the right atrium so it can be pumped into the pulmonary circuit.

keratin – a class of proteins that forms part of skin and nails (and animal horns and antlers etc.). These proteins are known for their toughness.

kidney – an oblong, bean-shaped organ in the dorsal, posterior region of the abdominal cavity. Supplied with blood from the renal artery, a kidney contains the nephrons, which remove and concentrate urine for its excretion.

kinetic energy – energy of motion. The kinetic molecular theory states that matter is made up of particles that are in constant motion. As temperatures decease, this motion is reduced, i.e., there is less kinetic energy, and the likelihood of chemical reactions is reduced. This explains the decrease in enzymatic activity at temperatures lower than optimum.

labia – lips. There are two folds of skin (labia), an outer and an inner one that protect the opening of the vagina.

labor (labour) – muscle contractions of the myometrium associated with birth. The intensity of these contractions is magnified by the hormone oxytocin from the posterior pituitary gland, which has a positive feedback relationship between pressure and contractions. Labor pains end when birth occurs.

lacteal – the beginnings of the lymphatic system in the villi of the small intestine. The lacteals transport the products of fat digestion away from the intestines.

large intestine – (AKA colon) – the large intestine begins at the ileo-caecal valve, has four major sections: ascending, transverse, descending colons and the rectum. The appendix is a vestigial pouch extending from the caecum, the first lobe of the ascending colon. The large intestine functions to absorb water left over from the digestion processes. It also contains bacteria (E.coli) that aid in the formation of feces.

larynx – voice box. The larynx is the first actual structure of the trachea. It is located below the epiglottis. The larynx contains two flaps of skin called the vocal chords. The differing tension (tautness) of the vocal chords produces the different pitches used in vocalization (talking, singing, etc.)

leukocyte – white blood cell. There are several kinds of white blood cells all of which help combat infection. Some of the white blood cells are phagocytic and engulf foreign, potentially harmful cells (or viruses). Others release specific proteins such as antibodies, or histamines etc. See also: "antibody", "histamine", "phagocytosis".

lipase – lipid-digesting enzyme. Lipase is produced by the pancreas and enters the duodenum as a component of pancreatic juice. The products of its activity are fatty acids and glycerol. The efficiency of lipase is increased by the activity of bile. See also: "bile".

lipid – the class of biochemicals that are immiscible in water. Lipids are sometimes called macromolecules, but they are limited in the size of polymers they can form. Examples of lipids include fats, oils, phospholipids, and steroids.

lipoprotein – a specialized molecular combination of lipid and protein that lines the inner walls of the alveoli. Lipoproteins reduce the likelihood that alveoli could collapse and their inner walls stick together by reducing the surface tension in the alveoli.

liver – the largest inner organ of the body. The liver is located posterior to the diaphragm and has many functions including the production of bile, the conversion of ammonia into urea, storage of glucose as glycogen, production of globulins, and etc.

Lock and Key Analogy – This is the model that is used to help explain enzyme function. According to this model, enzymes have specialized regions of their surfaces called active sites. Substrate molecules fit into these active sits making an "activated enzyme-substrate complex", which causes a slight shape change (induces a fit) resulting in the formation of products. See also: "activated E-S complex", "induced fit model".

loop of Henle – the middle region of a nephron. The loop of Henle lies between the proximal and the distal convoluted tubules. The filtrate entering the loop of Henle has already had the useful materials removed from it (nutrient molecules, some water and salts), so it has a similar composition to very dilute urine. Starting at the cortex, the loop of Henle extends down into the medulla of a kidney. As it does it encounters a hypertonic environment (high salt concentration), which draws water out of the urine, thus concentrating it. The water returns to the blood, which was made hypertonic by pressure filtration. The loop then leaves the medulla and goes back to the cortical region. This potion is not permeable to water, but it is permeable to ionic movement. Because of the departure of the water the urine here has a greater salt concentration than the surrounding fluid, sodium and chlorine ions leave the urine. At the upper portions of the ascending portion of the loop, the ions are actually pumped out of the nephron. These ions create the high salt concentration in the extra-cellular fluids around the nephron.

lubricant – a substance that reduces friction. In the body, water is a lubricant as it exists throughout tissues enabling harmless, and abrasionless movement, which not only permits flexibility, but also efficient movement of joints, breathing, and so on. One of the most obvious cases of water as a lubricant is in swallowing, where it liquefies food for swallowing and digestion.

lumbar – the third lowest of the four regions of the spine (following cervical and thoracic). Sympathetic neurons of the ANS arise from the thoracic and lumbar regions of the spine.

lumen – the central space of a tubular structure. Blood travels through the lumen of blood vessels, sperm through the lumen of the vas deferens etc.

lungs – organs of gas exchange for the respiratory system. These organs are comprised of a total of five lobes (three on the right and two on the left, accommodating the larger left ventricle) located at the terminus of the bronchi and bronchioles. The functional structures of the lungs are the alveoli and the associated pulmonary capillaries. See also: "alveoli", "pulmonary capillaries".

luteal phase – the phase of the ovarian cycle that is dominated by the activity of the corpus luteum. During this phase, which typically lasts from day 14 to day 28, the corpus luteum produces and secretes progesterone, which maintains the endometrium, and it takes over the production of estrogen for the maintenance of 2° sex characteristics. The corpus luteum begins to degenerate after about 10 days causing a drop in these hormone levels followed by menstruation.

luteinizing hormone (LH) – this is the gonadotropic hormone that promotes hormone production and release by the gonads. It is produced at the anterior pituitary and secreted in response to the release of a GnRH from the hypothalamus. In males, it results in the release of testosterone. In females, it promotes the release of progesterone and estrogen. As estrogen levels in females increase, the LH (and FSH) release is suppressed until the estrogen level surpasses a threshold, which results in the LH surge (an example of positive feedback). The LH surge causes ovulation, which facilitates the establishment of the corpus luteum. See also: "gonadotropic hormone", "estrogen", "positive feedback".

lymph – the fluid of the lymphatic system. As such it is mostly water, which is collected from tissue spaces. Most of this water originates as leakage from the capillaries.

lymph capillaries – the minute beginnings of lymph ducts that are located in tissue spaces of the body. These capillaries absorb the fluids that become lymph. The lacteals are specialized lymph capillaries that are located in the villi of the ileum, where they absorb fats.

lymph duct – Vessels that transport lymph. Lymph capillaries deliver the lymph to these larger ducts, which transport it to the circulatory system. Lymph ducts have valves much the same as veins as there is no pressure forcing along the movement of lymph. The movement is dependent upon the movement of the surrounding muscle tissue.

lymph node – a spot along lymph ducts, usually where two or more ducts join together. These nodes house white blood cells, which are active in the destruction of foreign and potentially harmful cells and substances transported by the lymph before it is added to the circulatory system.

lymphatic system – The system that is made up of all the vessels and nodes that transport and clean lymph. The lymph ducts all eventually join together to produce the main lymphatic ducts that join the circulatory system. The junction of the lymphatic system and the circulatory system is close to where the right subclavian vein meets the superior vena cava.

lysis – bursting. The bursting of red blood cells due to increased turgor pressure is called hemolysis. Plasmolysis is where water is drawn out of cells due to a hypertonic environment. See also: "hemolysis", "plasmolysis".

lysosome – an organelle produced by the Golgi bodies. Lysosomes contain hydrolytic enzymes and have the role of conducting intracellular digestion. Lysosomes have a double outer membrane, which helps prevent their rupturing. Some cells are autodigestive, which requires the rupturing of lysosomes. See also: "autodigestion", intracellular digestion".

maltase – the disaccharidase that catalyses the hydrolysis of maltase as well as the dehydration synthesis of glucose molecules to make maltose. Maltase is produced by the small intestine and is a component of intestinal juice.

maltose – the disaccharide that is made by the combining of two glucose molecules together. The digestion of starch (and glycogen) by amylase enzymes produces maltose. Maltose is further digested into glucose by maltase. See also: "maltase", "disaccharidase" AXY, F

matrix – the inner fluid of the mitochondria. The cristae membrane surrounds the matrix. It is structurally significant in the final stages of cellular respiration.

medium – background, o surrounding substance. For example, blood cells are suspended in a watery medium called plasma. The viscosity of the medium affects the rate of diffusion. See also: "viscosity".

medulla – The inner tissue of a organ or gland, as opposed to the outer cortex. Examples include the renal medulla, and adrenal medulla.

medulla oblongata – The elongate portion of the CNS at the base of the brain. Its location is also described as the brain stem. The medulla oblongata controls several reflexive, homeostatic actions involving non-skeletal muscle coordination, such as heart rate, breathing rate, swallowing, and vomiting.

meninges – the set of three membranous structures that surrounds and protects the CNS. The outermost layer, called the dura mater is the thickest and toughest. The inner layer is called the pia mater. Between these two is the arachnoid layer, which is specialized with blood vessels and capillaries. Inflammation of the meninges is known as meningitis.

menses – The phase of the uterine cycle lasting about five days during which menstruation occurs. See also: "menstruation", "uterine cycle".

menstrual cycle – part of the uterine cycle that focuses only on the period of menstruation vs. non-menstruation. See also: "menstruation", "uterine cycle".

menstruation – the sloughing off of blood tissue from the endometrium of the ovary brought on by low progesterone levels. Menstruation occurs during the menses phase of the uterine cycle and typically lasts about five days. See also: "menses", "uterine cycle" KXY

mesenteric artery – the blood vessel that conducts blood to the intestines.

messenger RNA – the single-stranded nucleotide polymer that is made by transcription from a segment of DNA known as a gene. MRNA carries the genetic information required for the synthesis of a specific protein from the nucleus to the ribosomes where translation occurs. See also "transcription".

metabolic pathway – a series of enzymatic reactions were the product of one step becomes the substrate of the next step. A different enzymes catalyzes each step in a metabolic pathway. Often, the product of a metabolic pathway will regulate its production by having a negative impact on the first step in its production. This is an example of negative feedback.

metabolic waste – waste products from cellular metabolic activity. Common metabolic wastes include CO_2, from cellular respiration, O_2, from photosynthesis, and NH_3, from the deamination of amino acids.

metabolism – the chemical activities of a cell that are controlled by enzymes.

microfilament – protein filaments that form part of the cytoskeleton of a cell.

microtubule – part of the cytoskeleton of a cell. Microtubules are made from the protein tubulin and are involved with cell movement. Cilia and flagella are examples of microtubular structures.

microvillus – membranous extensions of the cytoplasm of columnar epithelial cells that line the villi. The microvilli increase the contact surface area of the villi against the intestinal lumen, thus increasing the absorption of the products of digestion.

midpiece – the middle of the three main parts of a sperm cell, between the head and the tail. The midpiece is specialized by the presence of mitochondria that metabolize the fructose found in semen to produce the energy used by the tail for locomotion.

mitochondria – the organelle that conducts the metabolic reactions of cellular respiration. More precisely, cellular respiration involves the oxidation of carbon compounds to produce ATP. When the source of these carbon compounds is glucose (which is usual), the whole process is carbohydrate metabolism. Mitochondria have a convoluted inner membrane that is folded into shelf-like structures called cristae. The increase in membrane surface area created by the folding of this membrane increases the efficiency of the mitochondria at conducting cellular respiration.

mixed nerves – nerve tissue composed of bundles of long fibers of neurons (sensory dendrites and motor axons). In an organism, mixed nerves form nerve tracts and are visible as a shiny thin cord. The individual nerve cell fibres are protected from cross-communicating with each other by the Schwann cells, which make up the myelin sheath.

monoamine oxidase – the synaptic enzyme that destroys the neurotransmitter, norepinephrine. See also: "norepinephrine". I

monoglyceride – a neutral fat consisting of glycerol and one fatty acid. See also: "neutral fat".

monomer – (AKA unit molecule) the building block of a polymer. Monosaccharides (typically glucose) are the monomers of polysaccharides, fatty acids and glycerol are the monomers of neutral fats, amino acids are the monomers of proteins, and nucleotides are the monomers of nucleic acids.

monosaccharides – single sugars. Glucose, produced by photosynthesis, is the most common monosaccharide. Other examples of monosaccharides are fructose, galactose, ribose, deoxyribose, etc. Some monosaccharides can be combined to form disaccharides. Similarly, they can be produced by the hydrolysis of a disaccharide.

motor end plate – the terminus of a motor neuron's axon at an effector. Normally, axons end with a knob-like structure, but in this case, the terminus forms a flattened out structure. The same process of synaptic transmission occurs. See also: "axon", "synaptic transmission".

motor neuron – the type of nerve cell that has a long axon and a short dendrite, and whose cell body is located in the CNS. The axon conducts an impulse out of the CNS to an effector, (or in the case of the ANS, to a second motor neuron).

mouth – the anterior opening of the digestive tract.

mucosal cell – a mucus-producing cell. See also: "mucus".

mucus – (also spelled mucous) a watery fluid secreted by mucosal cells. Mucus is used in the body as a lubricant and a protectant (as in the case of gastric juice).

muscle – one of two types of effectors in the body (glands are the other). Muscles have the ability to contract (shorten) and thus confer movement. There are three types of muscle tissue: skeletal (somatic) – attached to the skeleton, cardiac – the heart muscle, and smooth – inner muscles of which we have little or no control, such as of the digestive system, and the diaphragm.

mutation - a change in the genetic structure of a chromosome. Mutations are caused by mutagens (agents of mutation), which fall into two categories: chemical and radiation. Mutations, themselves, fall into two categories: chromosome mutations, which involve changes to the overall structure of chromosomes; and gene or point mutations, which involve nucleotide changes within a gene. Addition, deletion, and substitution are three types of gene mutations. See also: "addition", "deletion", and "substitution".

myelin sheath – The fatty covering of motor axons and sensory dendrites outside of the CNS The covering is made out of Schwann cells. The myelin not only insulates the neural fibers from each other preventing cross-communication, but it also speeds up the transmission by causing saltatory transmission. See also: "saltatory transmission", "Schwann cell".

myelinated nerve fiber – a nerve fiber that is covered with myelin. See also: "myelin sheath".

myometrium – the muscle lining of the uterus (as opposed to the endometrium).

Na/K pump (sodium/potassium pump) – This is a membrane protein that pumps sodium ions one-way across it while pumping potassium ions the other way. It is called a pump because it requires ATP as neither ion is being moved down a concentration gradient. The Na/K pump is responsible for establishing and maintaining the resting potential of nerve cell membranes.

NAD (nicotinamide adenine nucleotide) – This is a molecular complex (coenzyme) that gains and transports hydrogen atoms from reactions. In doing so, it gets "reduced" and forms $NADH_2$. The hydrogen atoms are energy-rich and are subsequently used in another metabolic step that requires energy, such as cellular respiration and photosynthesis. NAD contains the vitamin niacin.

nasal cavity – The sinus chamber at the top of the nostrils. It is the main entry point for air into the respiratory pathway and is kept moist with mucus. It functions to trap debris that is in the inhaled air.

negative dipole – the opposite of a positive dipole. When electrons are shared unequally, as with a polar covalent bond, the associating atoms gain a slight charge. This charge is positive, if they have lost the use of one or more electrons or negative, if they have gained the use of one or more electrons. For example, the oxygen atom in a water molecule has a negative dipole, where the hydrogen atoms have a positive dipole. The symbol used to represent a negative dipole is δ. See also: "dipole", "positive dipole".

negative feedback – the interaction between molecules where a product molecule reduces or regulates the rate that it is being produced. An example of negative feedback using hormones is the regulation of testosterone levels in males. As the level increases, it affects the hypothalamus and anterior pituitary gland to decrease the production of LH and, subsequently, more testosterone. A decreased testosterone level has the opposite effect. In this manner, the level of testosterone is maintained fairly steady.

nephron – the functional unit of a kidney. Each kidney contains millions of nephrons. A nephron is a microscopic tubular structure that begins with Bowman's capsule, which encloses tuft of blood capillaries called a glomerulus. Other specialized regions of a nephron are the proximal and distal convoluted tubules, the loop of Henle and the collecting duct, all of which are surrounded by a peritubular capillary network. The combined function of these structures is the production of urine.

nerve – a combination of nerve cell fibers. See also: "mixed nerve".

nerve tract – a mass of nerves. The white matter of the spinal cord contains many nerve tracts that run up and down the CNS taking impulses to the brain for integration and messages down the spinal cord for transmission to effectors.

neuroendocrine control center – This is the combination of the hypothalamus (nerve tissue) and the pituitary glands (endocrine tissue). The hypothalamus is sensitive to blood conditions and responds by triggering the release of specific hormones from the pituitary glands. In the case of the posterior pituitary, the hypothalamus makes oxytocin and ADH, which are delivered to the posterior pituitary gland through nerves. These hormones are differentially released by the posterior pituitary gland. The anterior pituitary, in contrast, synthesizes the hormones that it releases (LH, FSH, etc.). "Releasing hormones" from the hypothalamus govern the release of the hormones from the anterior pituitary. See also: "anterior pituitary", "posterior pituitary", and "hypothalamus".

neuron – nerve cell. There are three categories of nerve cells: motor neurons, sensory neurons, and interneurons. Every nerve cell has the same three parts: dendrite, cell body, and axon.

neurotransmitter – the chemical substance (often a peptide molecule) that is released by an axon for the sole purpose of affecting the cells on the other side of a synaptic gap. Receptor sites on postsynaptic membranes receive the neurotransmitters. They impact its permeability, thus completing the transmission. Enzymes present in synaptic gaps destroy neurotransmitters so that the transmission can be repeated. Examples of neurotransmitters are noradrenaline and acetylcholine.

neutral fat – the combination of fatty acids and alcohols such as glycerol. Glycerol-based neutral fats can be monoglycerides, diglycerides, or triglycerides, depending on the number of fatty acids involved.

nitrogenous base – the bases in nucleotides all contain nitrogen, and are therefore referred to as nitrogenous bases. There are two groups of nitrogenous bases: purines and pyrimidines. See also: "purine", "pyrimidine".

nitrogenous waste – metabolic waste product containing nitrogen. Cells produce ammonia, NH_3, a toxic waste product through the deamination of amino acids.

nodes of Ranvier – the exposed spots along neurons between Schwann cells making up the myelin sheath. The movement of ions resulting in impulse transmission only occurs at the nodes of myelin-covered neuron fibers. This effect is called saltatory transmission. See also: "Schwann cell", and "saltatory transmission".

noncompetitive inhibitor – a chemical substance that prevents an enzyme from functioning by affecting its shape. Their mode of action is in contrast to competitive inhibitors that will occupy an active site to prevent the enzyme from functioning. Non-competitive inhibitors often work by bonding onto the enzyme in some spot other than the active site, which promotes a shape change to the enzymes. Working in a manner similar to heavy metal ions, they cause the active site to change shape and prevent their association with he substrate. See also: competitive inhibitor.

non-polar – refers to molecules and molecules bonds where the sharing of electrons is relatively equal, thus not creating any polarity.

noradrenaline (AKA: norepinephrine) – the neurotransmitter of the sympathetic nervous system. Noradrenaline is an excitatory neurotransmitter for many effects, with the exception of the various parts of the digestive system. Monoamine oxidase is the enzyme that destroys noradrenaline.

norepinephrine – See noradrenaline.

nose hair – hair in the nasal passageway that help catch debris and prevent its entry into the rest of the respiratory passageways.

nostrils – the nasal openings.

nuclear membrane (envelope) – the double membrane containing pores, which surrounds a nucleus. The nuclear membrane controls the passage of materials into and out of a nucleus.

nuclear pore – a protein-bound opening through the nuclear membrane where large substances are moved in or out of the nucleus. (Enzymes get moved in; ribosome subunits out, etc.).

nuclease – a nucleic acid-digesting enzyme. Nucleases are components of both pancreatic juice and intestinal juice.

nucleic acid – the class of polymers made out of nucleotides; either DNA or RNA.

nucleolus – the region of a nucleus containing the chromatin that is associated with making the ribosomal subunits. A nucleolus is visible because of its specialize chemical composition. It does not have a membrane.

nucleoplasm – the fluid of the nucleus. Chromatin (DNA) is suspended in the nucleoplasm. Chemically, the nucleoplasm and the cytoplasm are very similar.

nucleotide – the monomer of nucleic acids. Each nucleotide consists of the phosphate group, a 5-carbon sugar (ribose or deoxyribose), and a nitrogenous base.

nucleus – the region of a cell surrounded by the nuclear membrane. This region contains the cell's genetic material, and is therefore the locations of replication, transcription, as wells a where gene mutations occur.

null hypothesis – a hypothesis that is not supported by data. A null hypothesis (a incorrect educated guess) must be rejected, and should lead to further investigation.

observations – qualitative and quantitative data that is collected during an investigation. Interpretation and analysis of the observations allows one to make a conclusion.

oils – unsaturated fatty acids which are generally liquid at room temperature.

oligosaccharide – a sugar chain consisting of a few sugar molecules ("oligo" = few). Oligosacchardes on phospholipids make glycolipids; oligosaccharides on membrane proteins make glycoproteins. Both of these structures can be found on the outer membrane surface of animal cells and contribute to the cell recognition function. See also: "glycoprotein", "glycolipid".

oogenesis – the production of ova. Oogenesis begins as an embryonic process and finishes cell by cell during the ovarian cycle. See also: "follicle", "ovarian cycle".

optimum – the best. Enzymes function most effectively at optimum conditions because their active sites most closely match the shape required for their substrates. Deviations from the optimum conditions (i.e., temperature and/or pH changes) are less than ideal for enzymatic activity because they cause changes in the shape of the enzyme's active site.

organelle – a functional structure in a cell. Organelles include: ribosomes, Golgi bodies etc.

organic – containing carbon. Molecules that contain carbon are called organic as opposed to those without carbon, which are inorganic.

osmosis – the movement of water according to the Laws of Diffusion. Water molecules (as with solute particles) are at the mercy of concentration differences and (when possible) will move to even out these concentration differences. The "diffusion of water" as osmosis is often called occurs in a direction towards higher solute concentrations (away from higher water concentrations), or down an osmotic gradient. Osmosis from cells causes effects like plasmolysis and so on.

osmotic gradient – the difference in water concentration between two regions. See also: "osmosis".

osmotic pressure – The force water exerts on a membrane due to concentration differences. Osmotic pressure is critical to survival as it returns water back to the circulatory system during capillary tissue-fluid exchange. An even more specialized example of this can be found with nephron function. See also: "osmosis", "capillary tissue-fluid exchange", and "loop of Henle".

ova – female gametes (eggs). Ova are produced in ovaries by oogenesis. See also: "oogenesis", "follicle", and "ovarian cycle".

oval opening – a valve allowing blood to pass from the right ventricle of a fetal heart into the left ventricle, thus the blood bypasses the non-functional lungs. Once the lungs inflate (first breath), a significant volume of blood enters the left ventricle via the pulmonary veins. This forces the valve closed, and in a few days the tissues of the valve knit together, sealing the valve shut. See also: "fetal circulation".

ovarian cycle – the series of changes that occur in the ovaries on a 28-day cycle. The complete cycle has three phases: follicular, ovulation, and luteal.

ovary – female gonad. Oogenesis occurs in the ovaries as directed by follicle stimulating hormone (a gonadotropic hormone). See also: "oogenesis", "ovarian cycle", and "follicle stimulating hormone".

oviduct – See: fallopian tube.

ovulation – the bursting of a mature (Graafian) follicle, which is brought about by the LH surge on about the 14th day of the 28-day ovarian cycle. The result of this is the release of an ovum from the ovary. The ovulatory phase is the middle phase of the ovarian cycle (between the follicular phase and the luteal phase.

ovum – singular of ova. See: "ova".

oxygen – a diatomic gas found in nature making up about 21% of the atmosphere. Oxygen is required by mitochondria for cellular respiration and is produced by chloroplasts during photosynthesis.

oxygenated – with oxygen, as opposed to deoxygenated. Oxygen enters the blood by diffusion during external respiration, and is transported to body cells by hemoglobin in the form of oxyhemoglobin (HbO_2). It enters extracellular fluids during capillary tissue-fluid exchange, after which it diffuses into cells (internal respiration).

oxyhemoglobin – the combination of oxygen and hemoglobin (HbO_2). Each hemoglobin molecule has four bonding sites for oxygen.

oxytocin – a short protein hormone (9 amino acids long). Its production by the hypothalamus and its release by the posterior pituitary gland of females cause the contractions of the uterus known as labor pains. When a baby is about to be born, it applies pressure on the cervix, resulting in these impulses that signal the release of more oxytocin. This is an example of a positive feedback mechanism. See: "positive feedback mechanism".

pacemaker – See: SA node.

pancreas – the gland posterior to the stomach that has both endocrine and exocrine functions. As an endocrine gland, its islets of Langerhans produce the hormones insulin and glucagon. As an exocrine gland, it produces pancreatic juice. See also: "islets of Langerhans", "pancreatic juice".

pancreatic amylase – the component of pancreatic juice that digests starch (and glycogen) into maltose.

pancreatic duct – the tube through which pancreatic juice enters the duodenum.

pancreatic juice – the exocrine secretion from the pancreas. Pancreatic juice contains pancreatic amylase, trypsin, lipase, nucleases, sodium bicarbonate, and water. Water is used as a lubricant and as an ingredient in hydrolysis reactions. The bicarbonate ions buffer the pH to about 8.2 to facilitate the activity of the other components of pancreatic juice (which are all enzymes) and the enzymes of intestinal juice.

parasympathetic division – a subset of the autonomic nervous system (as opposed to the sympathetic division). The parasympathetic nerves consist of motor neurons that release acetylcholine to affect the effector. Acetylcholine will slow down the heart, and reduce breathing rate yet increases digestive activity. The nerves leave the CNS from the sacral and cranial regions. The parasympathetic effector system requires two motor neurons to reach the effectors, meaning there are parasympathetic ganglia (nerve cell bodies) outside of the CNS. These are distally located, near the effectors.

passive transport process – transport process that satisfies the Laws of Diffusion, therefore, does not involve the use of energy. Diffusion, osmosis, and facilitated diffusion are all passive transport processes.

pelvic region – See: "renal pelvis".

penicillin – an antibiotic. Penicillin prevents bacterial cell division. Antibiotics like penicillin gives one's immune system an upper hand in combating infection. Penicillin is excreted into the distal convoluted tubules of the nephrons in the kidneys (tubular secretion). J

penis – Male copulatory and excretory organ. The penis contains erectile tissue that will make it stiff when engorged with blood. Unlike excretion, copulation is not possible unless the penis is stiff.

pepsin – a protease enzyme. Pepsinogen (inactive form) is a component of gastric juice. HCl, another component of gastric juice, activates it. Pepsin breaks some of the peptide bonds holding specific amino acids together producing "polypeptides", which are sections of proteins of various lengths.

pepsinogen – See: pepsin.

peptidase – an enzyme that breaks peptide bonds. Peptides of various sorts are components of intestinal juice. Their substrates are dipeptides that have withstood the action of protease enzymes pepsin and trypsin. The products of their hydrolytic activity are amino acids.

peptide bond – a strong covalent bond that forms as a result of the dehydration synthesis of amino acids. The bond links the carboxyl end of an amino acid to the amino end of another amino acid.

peripheral nervous system – the portions of the nervous system that are not central. The axons of motor neurons (including the ANS motor neurons), and the cell body and dendrite of sensory neurons are generally considered to in the PNS. As such, the PNS is the link from the receptors to the CNS and from the CNS to the effectors.

peristalsis – smooth muscle contractions long tubular structures. It is the result of the coordinated contractions of both longitudinal and citcular muscles. It provides the movement of materials along soft-walled tubes that would normally be closed. It accounts for swallowing and for the movement of material all along the intestinal tract. Peristaltic action also moves urine along the ureter from the renal pelvis to the urinary bladder.

peritubular capillaries – the capillary bed that surrounds the a nephron of the kidneys. A kidney has millions of nephrons, likewise, it would have millions of peritubular capillary beds. The blood in these capillaries starts out very hypertonic due to pressure filtration through the Bowman's capsule. They contribute to urine formation by accepting the nutrients that are pumped out of the proximal convolute tubule and the water that is drawn out by the hypertonic conditions in the medulla. At the distal portion of the nephron, the composition of the blood in the capillary bed is adjusted (pH, presence of histamines, penicillin, etc.) by tubular excretion/secretion.

permeability – The nature of a membrane to allow or disallow specific substances across it is called selective permeability. The permeability of some membranes can be changed by the presence of factors in their environment. These factors may affect the shape of specific proteins, such as the sodium and potassium gates in neuron membranes and the water

channels in the collecting duct. See also: "selectively permeability", "ADH".

pH – a measure of the amount of hydrogen ions in a solution. If a solution is neutral, it has a pH of 7.0. Mathematically, a neutral solution has a hydrogen ion concentration ($[H^{1+}]$) of 1×10^{-7}. If there are more hydrogen ions than this, the pH is lower, meaning the solution is acidic. pH can be calculated as the $-\log [H^{1+}]$. See also: "acid", "base".

phagocytosis – endocytosis of large particles, such as foreign cells by white blood cells. Phagoctosis involves the use of cytoplasmic extensions called pseudopodia and results in the formation of food vacuoles that fuse with lysosomes for intracellular digestion. Phagocytosis is sometimes called "cell-eating", in contrast to cell drinking (pinocytosis). See also: "endocytosis", "pinocytosis".

pharynx – the region at the back of the mouth where both food and air travel. The base of the pharynx is specialized by having openings to both the esophagus (for food) and the trachea (for air). The epiglottis protects the opening to the trachea during swallowing.

phosphate – the ion formed from the oxidation of phosphoric acid. Chemically, it is PO_4^{3-}, and it is sometimes referred to as "inorganic phosphate, P_i". Phosphates are significant in phospholipids, nucleotides and ATP. See also: "phospholipid, "nucleotide", "ATP".

phospholipid – a neutral fat consisting of glycerol with two fatty acids bonded to it. The third bonding site is occupied by phosphate, which in turn has the N-containing choline group bonded to it. The effect of this is to create a fairly large molecule that has both a non-polar end (fatty acids) and polar end, where the P_i has a negative dipole and the N has a positive dipole. The significance of phopholipids is immediately apparent in the structure of membranes. See also: "fluid mosaic model".

phospholipid bilayer – A double layer of phospholipids arranged so that their polar ends are apart from each other and their non-polar regions are together, thus providing an oily nature to the inner region. See also: "fluid mosaic model".

phosphoric acid – H_3PO_4. See: "phosphate group".

phosphorylation – adding a phosphorous group to a molecule. Normally, with the phosphorous goes the extra energy. In this manner molecules are "energized" for activity. For example, glucose is phosphorylated at the beginning of cellular respiration; ADP is phosphorylated into ATP.

photosynthesis – Electrons from chlorophyll are easily "excited" by light energy, temporarily providing them with additional energy. This energy is utilized to build glucose molecules in the process of photosynthesis as the electrons are returned to their normal energy level. This process requires carbon dioxide and water and produces oxygen that must be released for the plant cells. The products of photosynthesis are the reactants in cellular respiration.

physical digestion – the mechanical breaking apart of food materials in the digestive system. Physical digestion occurs in the mouth (chewing, or mastication), in the stomach (as a result of churning), and in the duodenum (the emulsification of lipids by bile). See also: "emulsification".

pinocytosis – endocytosis of small particles. Endocytosis involves the invagination of the cell membrane to form a "pit" into which materials to be taken in move. Subsequently, the pit pinches off forming a vesicle, which can move to a lysosome for the digestion of its contents. Pinocytosis is sometimes referred to as "cell drinking", in contrast to cell eating (phogocytosis). See also: "endocytosis", "phogocytosis".

pituitary gland – the endocrine gland attached to the hypothalamus. There is an anterior lobe and a posterior lobe of the pituitary gland. The anterior lobe secretes a variety of hormones including LH and FSH when stimulated to by releasing hormones from the hypothalamus. The posterior pituitary secreted oxytocin and ADH, which are produced by the hypothalamus. See also: "neuroendocrine control center", "anterior pituitary gland", and "posterior pituitary gland".

placenta – the exchange tissue consisting of the interface between maternal blood supply in the endometrium and fetal blood supply at the distal terminus of the umbilical cord. It is through this tissue that the embryo (later called the fetus) receives its nutrients and oxygen, and gets rid of wastes such as CO_2 and urea. In its early stages of development, it produces HCG to maintain the corpus luteum (producing and releasing progesterone and estrogen). When the placenta is well established, it takes on this role itself, thus preventing the breakdown of the endometrium.

plaque – hard deposits. Plaques form on teeth as a result of bacterial growth. They also form in the lumen of blood vessels as a result of fatty deposits. Here, they lead to partial blockage of the arteries (atherosclerosis), high blood pressure, and hardening of the arteries (arteriosclerosis). See also: "blood pressure".

plasma – the fluid portion of blood. Plasma is mostly water (approximately 91%), contains globulins (proteins from the liver), nutrients, wastes, cholesterol, steroid hormones, dissolved gases, and various ions.

plasma membrane – See: "cell membrane".

plasmolysis – the loss of cytoplasmic water due to being in a hypertonic environment. The effect of plasmolysis is cell shrinking. In plant cells, the cell membrane shrivels away from the cell wall as it loses turgor pressure. The opposite of plasmolysis is deplasmolysis. See also: "deplasmolysis", "crenation".

platelet – a type of blood cell formed by the fragmentation of a megakaryotype (a large cell from bone marrow). Platelets are also known as thrombocytes. These cells function for blood clotting. When damaged, a platelet releases thromboplastin, an enzyme that begins a cascade of reactions that ends with the formation of a blood clot. See also: "blood clot", "thromboplastin".

pleural – relating to breathing. See also: "pleural membranes".

pleural membranes – the double layer of membranes that surround the lungs. The outer pleural membrane is on the inner wall of the thoracic cavity and along the top of the diaphragm. The inner pleural membrane is on the surface of the lung tissue. During inhalation and exhalation, these two surfaces do not move together, yet allow the surfaces to slide over one another in a frictionless manner. The pleural membranes further contribute to inhalation. As the chest cavity is expanded, the outer membrane moves with the ribcage. The interpleural space (between the membranes) is moist and the movement pulls the inner membrane with it due to cohesion. If these membranes and the interpleural space do not remain intact, a pneumothorax (lung collapse) can result. In this regard, the pleural membranes are said to individually seal each lungs and its collapse.

pneumothorax – lung collapse. See: "pleural membranes".

polar covalent – a covalent bond where the electrons are not shared equally by the atoms involved. The bonds between H and O within a water molecule are polar covalent bonds. Dipoles are associated with polar covalent bonds. See also: "dipole", "negative dipole", and "positive dipole".

polar molecule – a molecule that has dipoles, such as water. Some amino acids are polar, as is the head of a phospholipid, thus allowing them to mix with water. Non-polar molecules, like lipids, will not mix with water.

polar surface – a surface that is hydrophilic (likes water). Membranes have polar surfaces because of the arrangement of their phopholipids. See also: "fluid mosaic model".

polarity – having either a positive or a negative dipole. See also: "dipole", "negative dipole", and "positive dipole".

polymer – a macromolecule made of monomers. For example, starch, glycogen, and cellulose are polymers of glucose, nucleic acids are polymers of nucleotides, and proteins are polymers of amino acids. Lipids do not form true polymers due to the limited number of monomers that can be incorporated into a single large molecule.

polypeptide – a large molecule made of many amino acids. Proteins are polypeptides. See also: "polymer", "primary structure".

polysaccharide – a large molecule consisting of many monosaccharides. Starch is a polysaccharide. See also: "polymer".

polysome – an organelle consisting of several to many ribosomes clustered together for the mass production of a protein. In this scenario, a succession of ribosomes simultaneously reads a single mRNA strand.

positive dipole - the opposite of negative dipole. When electrons are shared unequally, as in a polar covalent bond, the associating atoms gain a slight charge. This charge is positive, if there is a loss of the use of one or more electrons or negative. For example, the oxygen atom in a water molecule has a negative dipole, leaving the hydrogen atoms with positive dipoles. The symbol used to represent a positive dipole is δ^+. See also: "dipole", "negative dipole".

positive feedback mechanism – a biochemical interaction that promotes itself. Unless interactions like these meet with some outcome, they would be destructive. Here are two examples of positive feedback: (1) with reference to oxytocin, where childbirth is the outcome, and (2) with reference to estrogen's influence on LH, where ovulation is the outcome.

posterior pituitary – a portion of the pituitary gland. The posterior pituitary functions as part of the neuroendocrine control of fluid volume and childbirth because it releases antidiuretic hormone and oxytocin, respectively. See also: "neuroendocrine control", "antidiuretic hormone", and "oxytocin".

posterior vena cava – The major vein that drains the blood from the body regions posterior to the heart. It is also called the inferior vena cava, as opposed to the anterior, or superior, vena cava. All the venules in the lower regions of the body contribute blood to the posterior vena cava, which ends at the right atrium.

post-synaptic membrane – in synaptic transmission, this is the membrane that is equipped with receptor sites (proteins) to receive neurotransmitters.

potassium gate – a protein in the membrane of neurons that changes shape and allows potassium ions out of the neurons once the axoplasm gains a positive charge due to the influx of sodium ions. See also: "action potential".

precursor – an inactive form of a molecule. Protease enzymes (pepsin and trypsin) are produced as precursors to avoid autodigetion of the cell. They are activated once they are released. HCl activates pepsin in the stomach, and trypsin is activated by the enzyme enterokinase. See also: "protease", "pepsin", and "trypsin".

prediction – an estimation of a future outcome. A prediction follows a hypothesis in a scientific investigation.

preganglionic fiber – With reference to the fact that the effector system of the ANS is a two motor neuron system, there are ganglia (nerve cell bodies where synaptic transmission occurs) outside of the CNS. A preganglionic fiber refers to the axon between the CNS and the ganglion. Sympathetic neurons have short preganglionic fibers. In contrast, parasympathetic neurons have long preganglionic fibers.

pregnancy – ("with child". A pregnancy begins with the implantation of a growing embryo into the endometrium. With successful implantation, a placenta will develop and normal embryonic and fetal development will occur. Pregnancy ends with the secretions of oxytocin followed by birth. See: also "implantation", "placenta", and "oxytocin".

pressure filtration – the filtration process that occurs at the interface of the glomerulus and Bowman's capsule. Pressure filtration results in the formation of a filtrate that is mostly water, but also contains small molecules that are forced out of the blood and into the porous end of the nephron. Pressure filtration also results in hypertonic blood that leaves the glomerulus in the efferent arteriole. If blood pressure decreases, so does pressure filtration.

pressure gradient – the difference in pressure between two regions. Osmosis occurs due to osmotic pressure gradients. Gases diffuse during external respiration because of gas pressure gradients. See also "osmotic pressure", "external respiration".

pressure receptor – a nerve receptor that is sensitive to pressure. Pressure receptors exist in the carotid and aortic bodies. They respond to low blood pressure and cause the heart rate to increase.

pre-synaptic membrane – the membrane on the end of an axon at a synapse. During synaptic transmission, neurotransmitters are released from vesicles produced behind the pre-synaptic membrane at an axon terminus.

primary structure – the first of four levels of protein structure. Primary structure refers to the sequence of amino acids held together by peptide bonds. Primary structure is dictated by the nucleotide sequence in a DNA gene segment.

procedure – the methods followed in a scientific investigation.

progesterone – a female steroid hormone that is produced by the ovary, and specifically the corpus luteum following ovulation. Progesterone promotes the maturation of the glandular component of the endometrium making it secretory, as well affecting the milk-producing glands of the breasts in preparation for pregnancy.

prokaryotic – ("before kernal"). Karyon refers to "nucleus". Prokaryotic cells are simpler than eukaryotic cells, which have a true nucleus. Bacterial cells are examples of prokaryotic cells and, as such, lack all internal membranes, including a nuclear membrane. (Therefore, bacteria have no nucleus, or any other membrane-bound organelle). See also: "eukaryotic".

proliferative phase – part of the uterine cycle involving the thickening of the endometrium as a result of increased blood supply and the building up (proliferation) of the endometrial tissue under the influence of estrogen. The proliferative phase follows menstruation and typically lasts about nine days. See also: "uterine cycle", "estrogen".

prostaglandins – non-steroid, lipid-based chemical messengers of the body. Many tissues of the body produce and secrete prostaglandins. In the reproductive system, the prostate gland releases prostaglandins, which become a component of semen. These prostaglandins promote contractions of the uterus, which are believed to aid in the movement of sperm towards the fallopian tubes.

prostate gland – a male reproductive gland which secretes a mucus fluid that is rich with bicarbonate ions to buffer the semen against pH changes once introduced into the acid environment of the female reproductive system.

protease – a protein-digesting enzyme. Typically, to avoid their production in a cell, proteases are produced in an inactive (precursor) form. Pepsinogen and trypsinogen are proteases. See also: "pepsin", "trypsin", and "precursor".

protein – a polymer of amino acids; a polypeptide with a specific function. There are many types of proteins. Some are enzymes, others are hormones, antibodies, form cytoskeletal frameworks and/or the microtubules of cilia and flagella used for cell movement. All proteins have at least two of the four possible phases (or levels) of structure.

prothrombin – a globulin (liver-produced protein) that is soluble in plasma and contributes to the formation of a blood clot. Prothrombin is converted to thrombin by thromboplastin, which is an enzyme released by damaged platelets. This reaction requires to the co-factor Ca^{2+}. See also: "thrombin", "thromboplastin".

protoplasm – the living portion of cell. Generally, it includes the cell membrane and everything inside of it. A plant cell wall is not part of its protoplasm.

protractile protein – a cytoskeletal feature comprised of proteins which display secondary structure. When these proteins associate with specific chemical substances their shape changes and they elongate thus move organelles within the cytoplasm.See also: "contractile protein".

proximal convoluted tubule – the part of the nephron between Bowman's capsule and the loop of Henle. The proximal convoluted tubule is specialized to allow the selective reabsorption of nutrient molecules and other specific components of the filtrate that are still useful to the body. These substances are moved back into the peritubular capillaries. See also: "selective reabsorption", "peritubular capillaries".

ptyalin – See: salivary amylase.

pulmonary – pertaining to the lungs.

pulmonary artery/vein – The right and left pulmonary arteries conduct blood to the right and left lungs, respectively. They are the branches of the pulmonary trunk, which conducts blood out of the right ventricle. The pulmonary veins conduct blood from the lungs to the left atrium.

pulmonary capillary – a capillary that is associated with an alveolus. During external respiration carbon dioxide and some water leave the pulmonary capillaries and oxygen diffuses from the interior of the alveoli through the walls of the alveoli and into pulmonary capillaries where it bonds onto hemoglobin, producing oxyhemoglobin (HbO_2). The HbO_2 is then transported to systemic tissue capillaries.

pulmonary circuit / circulation - the portions of the circulatory system that relate to the lungs. The pulmonary circuit begins at the right ventricle, which pumps blood out the pulmonary trunk and into the two pulmonary arteries. It includes the pulmonary capillaries, where external respiration takes place, and the pulmonary veins, which conduct oxygenated blood to the left atrium.

pulmonary trunk – the artery that leaves the right ventricle. It branches to become the right and left pulmonary arteries.

pulmonary valve – the semilunar valve at the beginning of the pulmonary trunk. It functions to prevent the backflow of blood from the pulmonary trunk into the right ventricle.

purine – one of two categories of nitrogenous bases. The purines are large, double-ringed molecules. Adenine and guanine are purines. See also: "pyrimidine", "base", "adenine", and "guanine".

Purkinje fibers – These are nerve tracts that begin at the AV node in the right atrium, extend down the septum of the heart and out into the massive walls of the ventricles. The ventricular contractions are coordinated because of the simultaneous delivery of impulses by these nerves.

pyloric sphincter – This is the muscular constriction (valve) at the bottom of the stomach. The pyloric sphincter relaxes to permit small amounts of acid chyme into the duodenum at a time, so it can be neutralized and digested further.

pyrimidine – one of two categories of nitrogenous bases. The pyrimidines are small, single-ringed molecules. Cytosine and thymine are pyrimidines (as is uracil, in RNA). See also: "purine", "base", "adenine", and "guanine".

qualitative – describing "quality" as in qualitative observation, as opposed to quantitative observation.

quantitative – describing "quantity" as in quantitative observation, as opposed to qualitative observation. Quantitative observations contain numbers.

quaternary structure – the fourth level of protein structure. Not all proteins have quaternary structure, which is the association of tertiary-structured proteins. Hemoglobin is an example of a protein that has quaternary structure, as it is the association together of four different proteins around a central iron-containing heme group. See also: "hemoglobin".

radiation – part of the electromagnetic spectrum. Some forms of radiant energy (UV light, X-rays and such) are known to cause genetic mutations at certain short wavelengths and intensities. See also: "mutagen".

reabsorption of H_2O – The reabsorption of water is a key process that occurs in the circulatory and the urinary system. In both cases, an osmotic gradient is established through some sort of filtration process and water moves passively according to the osmotic gradient. See also: "capillary tissue-fluid exchange", "peritubular capillaries".

receptor – a protein in a cell membrane that has a bonding site for a substance that can be in its environment. Receptors exist on most cells. For example, target cells have receptors for protein hormones (e.g., insulin), and post-synaptic membranes have receptors for neurotransmitters. See also: "fluid mosaic membrane model".

receptor site – See: "receptor".

recombinant DNA – a segment of DNA that is constructed out of DNA from two sources. Normally, geneticists will enzymatically take a gene (segment of DNA) that specifies the production of a known protein and inset it into the genome of host cells. This alters the genotype of the host cells. Subsequent mass production of the cells will result in the mass production of the gene, and makes it possible to mass-produce the intended protein. Many protein pharmaceuticals (insulin, interferon etc.) are made in this way. Another advantage is that the hybrid cells can give an organism new properties (bacteria that can metabolize oil) and/or abilities like hybrid crops that are drought-resistant, etc.

rectum – the last portion of the large intestine. The rectum stores and compacts feces until defecation. See also: "large intestine".

red blood cell (AKA erythrocyte) – small cells that are mass produced by the red bone marrow. These cells are biconcave and anucleate (without a nucleus), thus they have a relatively short life span of about 120 days. Worn out and dead red blood cells are removed and destroyed by the liver. Red blood cells are significant because they contain hemoglobin, which can differentially transport oxygen, carbon dioxide, and hydrogen in the blood. See also: "hemoglobin".

reduced hemoglobin (HHb) – hemoglobin that is transporting hydrogen ions. Hydrogen ions are generated as a result of the carbonic anhydrase reaction that is associated with internal respiration. Hemoglobin that was transporting oxygen is now available to transport other substances. Some bonding sites pick up CO_2, making carbaminohemoglobin, where others pick up H^{1+}, making reduced hemoglobin. This uptake of H_{1+} helps to buffer blood by preventing a drop in pH. See also: "carbonic anhydrase".

reflex arc – the simplest complete neural pathway. A reflex arc typically involves a sensory neuron, an interneuron and a motor neuron. Stimuli are received, impulses are transmitted to the CNS and to a motor neuron for an immediate reaction. The advantage of a reflex arc is its immediacy of response. Additional spinal cord interneurons take impulses to the brain for interpretation, but this is not part of the reflex arc.

refractory period – the period of time following an erection during which the penis remains flaccid. See: also "erection".

releasing hormone (factor) – a chemical hormone-like substance that is produced by the hypothalamus for the purpose of affecting the anterior pituitary gland. The hypothalamus produces several different substances, which it puts into the short portal blood vessels to the anterior pituitary. Each substance stimulates the release of a different hormone from the anterior pituitary. Examples include the gonadotropic releasing hormones (GnRHs). See also: "gonadotropic releasing hormone".

reliability – with reference to the data from scientific experiments. If the experiment was conducted properly (controlled with appropriate sample sizes etc.), then the results can be considered to be reliable.

renal – refers to kidneys.

renal artery – the artery that is a branch of the aorta and conducts blood into a kidney. There is a right and a left renal artery.

renal cortex – the outside portion of a kidney. The renal cortex consists of glomeruli, afferent and efferent arterioles, Bowman's capsules and the convoluted tubules.

renal medulla – the inner tissue region of a kidney. The renal medulla is the location of the loops of Henle and the collecting ducts. The termini of the collecting ducts form the renal pyramids, which are extensions of the renal pelvis.

renal pelvis – the innermost chamber of a kidney. The renal pelvis collects urine from the various collecting ducts and retains it until it can be conducted to the urinary bladder by the ureter.

renal vein – the vein that conducts blood out of a kidney and into the posterior vena cava. There is a right and a left renal vein.

renature – conformation (shape) changes that return a protein to its shape during optimal conditions. If denaturing is not complete, some proteins (including enzymes) can renature. See also: "denature".

replication – the process during which DNA makes a copy of itself. This enzymatic process involves the "unzipping" of DNA by the enzyme DNA helicase, followed by the action of DNA polymerase enzymes which regulate the incorporation and bonding of new DNA nucleotides with the exposed bases on the parent strands. Once into place, these newly incorporated nucleotides bond together to make strands. This process occurs down the length of a DNA molecule and produces two identical strands (barring any mutations). Replication occurs prior to cell division to ensure that daughter cells can have the correct genotypes.

repolarization – the process that allows a neuron's membrane to regain the polarized condition it has during resting potential. Repolarization involves the opening of the potassium gates, which result from the sudden influx of sodium ions, which gives the axoplasm a positive charge (relative to the outside of the cell). The opening of these gates permits the outflow of potassium ions, which reduces this positive nature, thus regaining the resting potential. It is during this period that the Na/K pump is actively returning sodium and potassium ions to the correct sides of the membrane, this maintaining resting potential and making a subsequent impulse transmission possible. See also: "depolarization", "sodium/potassium pump".

RER – rough endoplasmic reticulum. This portion of the ER has ribosomes embedded in its membranes. These ribosomes make proteins that enter the ER to be packages into a vesicle for eventual secretion or to be used for some other metabolic purpose within the ER, such as the synthesis of steroids in SER regions.

respiratory – referring to "respiration".

respiratory center – the portion of the brain that monitors the blood and signals the contraction of the diaphragm and intercostals muscles causing inhalation. The respiratory center is located in the medulla oblongata. See also: "medulla oblongata".

respiratory tract – the pathway of airflow in the body associated with respiration. The respiratory tract includes the nasal passageways and sinuses, pharynx, larynx, trachea, bronchi, bronchioles, and finally the alveoli.

resting potential – the difference in electrical nature between the inner and outer surface of a neuromembrane when it is at rest (not conducting an impulse). This charge is brought about and maintained by the differential placement of ions. The resting potential is equal to –65 mV (inside relative to outside). Inside the neuron are potassium ions and negative ions. Outside the membrane is an abundance of sodium ions. Sodium/potassium pumps maintain the resting potential. Nerve impulses are disruptions to this polarized nature. See also: "impulse", Na/K pump.

R-group – the variable portion of an amino acid. There are 20 different R-group, one for each amino acid. The simplest R-group is a hydrogen atom (as in the case of glycine, one of the amino acids). The R-groups often have a specific chemical nature to them (polar, acidic, basic, etc.), which creates interesting chemistries for proteins and contributes to their 3° structures, the nature of active sites and receptor sites etc. See also: "amino acid", "tertiary structure".

ribonucleic acid (RNA) – the polymer of nucleotides that functions in the cytoplasm. There are three different types of RNA: mRNA, tRNA, and rRNA, all of which are involved in protein synthesis. RNA differs from DNA in that it has ribose sugar (not deoxyribose) and the base uracil (instead of thymine). See also: "mRNA", "tRNA", and "rRNA".

ribose – a pentose sugar ($C_5H_{10}O_5$). Ribose is the sugar that forms part of RNA nucleotides, and is therefore in RNA, as opposed to deoxyribose ($C_5H_{10}O_4$), which is in DNA. See also: "deoxyribose".

ribosomal RNA – one of the three types of RNA. rRNA is incorporated into ribosomal subunits as they are being made at the nucleolus. When these subunits form around a strand of mRNA, they help align the strand so translation can occur properly.

ribosome – a non-membranous organelle that has two subunits, which are assembled at the nucleolus. Each complete ribosome consists of a large subunit (with about 50 proteins) and a smaller subunit (about 30 proteins) The subunits leave the nucleus via nuclear pores and can exist free in the cytoplasm individually, or as part of a polysome, or they can become embedded into the walls of ER, (making the RER). All ribosomes conduct translation, the final stage of protein synthesis. See also: "translation".

RNA – See: "ribonucleic acid".

SA node – one of two nodes in the heart. The other is the AV node. These heart nodes are unique in that they are a combination of nervous tissue and muscle tissue, which allows them to initiate contractions. Both are located in the right atrium. The SA node is also called the pacemaker of the heart because it is responsible for the intrinsic (unassisted) heart rate of 72 beats per minute. It causes the contraction of the atria and sends an impulse to the AV node to stimulate its activity. The AV node sends impulses down the Purkinje fibers causing the contraction of the ventricles. The SA node is also connected to the medulla oblongata by both sympathetic and parasympathetic nerve fibers. The sympathetic fibers will increase its activity, where the parasympathetic will decrease its activity. See also: "AV node", "heart rate", and "ANS".

SA/V ratio – the relationship between the surface area of a cell and the volume of its cytoplasm. All nutrient-type substances that a cell requires and all secretory and excretory products of a cell must pass through the cell membrane. The surface area of a cell must be large enough to manage this rate of exchanges, but not so large that it is inefficient. As cells grow, their metabolic requirements increase, and the SA/V ratio decreases (metabolism occurs in the *volume* of the cytoplasm) and it can become increasingly difficult for the cell membrane to manage the required exchanges. This can either lead to cell division, cell death, a change in cell function, or no further growth.

saccule – a flattened sac. The Golgi apparatus is often described as a set of flattened saccules. See also: "Golgi apparatus".

sacral – the lowest region of the spinal cord. Some of the parasympathetic neurons arise from this section of the spinal cord.

saggital – a longitudinal section through a medial axis of a bilaterally symmetrical organism to create a right and left half.

salivary amylase – (AKA "ptyalin"). This is the starch- (and glycogen-) digesting enzyme that is a component of saliva. It has an optimum pH of about 7.0 in contrast to pancreatic amylase, which has an optimum pH of just over 8.0. As always, the product of amylase activity is maltose. Salivary amylase is denatured in the stomach and digested as with any other protein that is swallowed.

salivary gland – one of three sets of exocrine glands that deliver mucus (water) and salivary amylase to the mouth.

salivary juice (saliva) – a combination of salivary amylase and mucus (water) that is produced and delivered to the mouth through ducts from the salivary glands. See also: "salivary amylase", salivary gland".

saltatory transmission – saltatory means "jumping". Impulse transmission along myelinated fibers is described in this manner because of the positioning of Schwann cells. The ion movements associated with the impulse only occur at the exposed portions of the neuron – specifically the nodes of Ranvier, which are the spots between the Schwann cells. It is said that these impulses jump from node to node and can travel much faster in this manner. See also: "Schwann cell", "node of Ranvier".

sample size – refers to the number of trials in a scientific investigation. Larger sample sizes reduce the impact of individual differences. More reliable results can be obtained if large sample sizes are used.

saturated fatty acid – a fatty acid that does not contain any double bonds, and is therefore saturated with hydrogen atoms.

Schwann cell – a fatty cell first identified by Theodore Schwann in the 1800's. Schwann cells are considered to be specialized types of glial cell. Glial cells support nerve cells and nerve cell function. In particular, Schwann cells make up the myelin sheath and result in saltatory transmission. See also: "myelin sheath", "saltatory transmssion".

scientific method – an approved way of conducting a scientific investigation. The scientific method starts with a hypothesis that will give direction to the procedures that follow it. When set up with the appropriate controls and sample size, etc. the procedures will produce results that are valid and reliable. The results may or may not support the hypothesis. If they do not, then the hypothesis is called a null hypothesis. If they do, then a meaningful conclusion can be follow.

scrotum – the thin tissue that contains the testes outside of the body cavity and thus at a temperature lower than normal body temperature (34°C). A lower temperature is required for the normal development and maturation of sperm.

secondary structure (AKA 2° structure)– This refers to protein structure. All proteins are a sequence of amino acids (primary structure), but this sequence can take on either of two 2° conformations; an alpha helix (α-helix) or beta pleated sheet (β-pleated sheet), depending on specific bonding arrangements that form within the molecule. When an α–helix forms, H-bonds form between dipoles of successive loops of the molecule. When a β–pleated sheet forms, the H-bonds form between dipoles on adjacent sections of the molecule. A single protein can have sections of α–helix and β–pleated sheets.

secretion – the release of a cellular product (typically a protein or a lipid substance) that serves some purpose to the organism. For example, hormones and digestive enzymes get secreted from the cells that produce them. Secretion normally involves the joining of a secretory vesicle to a cell membrane.

secretory pathway – the route, as defined by a series of organelles, that secretory products follow as it is being prepared and secreted. The secretory pathway of a protein, such as an enzyme starts at the ribosomes, involves the RER, a transition vesicle, the Golgi apparatus, and finally a secretory vesicle, which fused to the cell membrane to release its contents through exocytosis. See also: "secretion", "secretory vesicle".

secretory phase – part of the uterine cycle. The secretory phase occurs during the second half of the uterine cycle and is marked by the maturation and function of mucous secreting glands in the endometrium. It typically lasts from day 15 to day 28. The secretory phase is the final step in preparing the uterus for implantation. It is brought about by high progesterone levels and ends with menstruation, when estrogen and progesterone levels drop. See also: "uterine cycle".

secretory vesicle – a vesicle that is prepared by the Golgi apparatus for secretion. Secretory vesicles normally contain either a protein substance (hormone, enzyme, neurotransmitter) or a lipid substance (steroid, oil) that has been made by the cell and will be used by the organism. See also: "secretion", "secretory pathway".

selective reabsorption – the process that occurs along the proximal convoluted tubule. During selective reabsorption, useful components of the filtrate that were forced through the porous Bowman's capsule are recovered into the peritubular capillary network. This is an energy-requiring process. See also: "proximal convoluted tubule".

selectively permeable – a feature of membranes that is determined by the specific proteins that are embedding in the phospholipid bilayer. These proteins equip the membrane for the transport of some substances, but not others through processes of facilitated or active transport. The acronym SPM refers to a selectively permeable membrane. See also: "fluid mosaic model".

semen – the combination of sperm and seminal fluid. This is the substance that leaves a penis during ejaculation.

semi-conservative – This refers to DNA structure. Each DNA strand produced (by replication) has one old (parental) strand, and one newly formed strand. This feature is called semi-conservative.

semi-lunar valve – the valves through which blood must pass to exit the heart. As such, there is a semilunar valve at the beginning of the aorta (called the aortic valve) and one at the beginning of the pulmonary artery (called the pulmonary valve). Semi-lunar valves prevent the backflow of blood into the ventricles.

seminal fluid – part of semen. Seminal fluid is a composite fluid from the seminal vesicle (which releases mucus), prostate gland (which releases buffers), and the Cowper's gland (which releases fructose). The seminal fluid helps to keep the sperm alive for a period of time after it is ejaculated.

seminal vesicles – one of three types of glands that contribute to the seminal fluid. The two seminal vesicles release mucus. See also: "seminal fluid".

seminiferous tubules – the tubules in the testes where spermatogenesis occurs. The sperm move from the tubules to the epididymis where they mature. See also: "epididymis".

sensory neuron – a type of nerve cell that conducts an impulse from a receptor to the CNS. Sensory neurons have long dendrites and short axons. Their cell bodies are located along dorsal root at ganglia.

septum – dividing wall. The heart has a septum that divides the ventricle chambers. In this case, the septum is the thick muscular wall of both chambers. The Purkinje fibers, which cause the ventricle contractions, run down the septum. Furthermore, the dividing wall between the nostrils is also called a septum. In this case it is made of cartilage. See also" "Purkinje fibers".

SER – smooth endoplasmic reticulum. This membranous organelle extends though the cytoplasm, often originating from the RER. The SER produces transition vesicles, which conduct cell products to the Golgi apparatus for preparation for secretion. These vesicles contain steroids, and some may also contain proteins made at the RER. The SER also contains enzymes that detoxify the cytoplasm of the cell. It additionally has a transport function in the cytoplasm since molecules move through it. See also: "secretory pathway", "RER".

simple sugar – See: "monosaccharide".

single sugar – See "simple sugar".

small intestine – the longest portion of the gastrointestinal tract lying between the stomach and the large intestine It is subdivided into three parts; the duodenum, where most of the digestion occurs, the jejunum where digestion finishes and absorption begins and the ileum, which is highly specialized for absorption. The small intestine is specialized for transport (circular and longitudinal muscles and its connections via sphincters to the stomach on one end and the large intestine at the other); digestion (ducts attach to the small intestine that delivery bile and pancreatic juice, as well as its own intestinal glands that secrete a variety of enzymes); and for absorption (though the specialized projections of the ileum called villi).

sodium bicarbonate – the buffering component of pancreatic juice. The bicarbonate ions maintain the pH of chyme at about 8.2, which is optimal for the function of the pancreatic enzymes. See also: "bicarbonate ion".

sodium gate - a protein in the membrane of neurons that changes shape and allows sodium ions into neurons once the neuromembrane is disturbed by stimulus that surpasses a threshold. See also: "action potential".

solubility – the ability of a substance to dissolve and form a solution in a solute.

solute – the portion of a solution that is in the lesser quantity. See also: "solution".

solution – the combination of a solute and a solvent. The portion that is in greater quantity is termed the solvent whereas the portion in lesser quantity is the solute. For example, with salt water, where salt is dissolved in the water, the salt is the solute and the water is the solvent.

solvent – the portion of a solution that is in greater quantity. See also: "solution".

somatic nervous system – the portion of the nervous system that conducts impulses to the skeletal muscles.

sorting center – See: "thalamus".

specific heat capacity – the ability of a substance to retain its temperature while gaining energy. Water has a high specific heat capacity – meaning that it takes a lot of energy to change its temperature. This property in water is a result of the hydrogen bonding that forms among the water molecules.

sperm – male gamete. Sperm are produced by spermatogenesis in the seminiferous tubules of the testes. Structurally, they have three main parts: a head, midpiece and a tail. The head is covered by an acrosome.

spermatogenesis – the process of cell division that leads to the production of sperm in the seminiferous tubules of the testes. After puberty, spermatogenesis is ongoing and millions of sperm are produced daily, unlike oogenesis where a single ovum develops in the ovary every 28 days.

sphincter muscle – a circular band of muscle that constricts the lumen of a tube. Arterioles and the gastrointestinal tract have sphincter muscles, which are used to regulate the flow of material through the lumen. See also: "arteriole", "cardiac sphincter", "pyloric sphincter", "ileo-caecal valve", and "anal sphincter".

spinal cord – the portion of the CNS that extends posteriorly from the base of the brain. It is protected by membranes called meninges and encased in the vertebral column. The spinal cord is divided into four regions: cervical, thoracic lumbar, and sacral. Thirty-one pairs of peripheral nerve originate along the length of the spinal cord. When viewed in cross-section, the differences between the white and gray matter are visible.

spinal nerve – a peripheral nerve that originates from the spinal column. A spinal nerve is a mixed nerve, as it contains both sensory and motor fibers. See also: "mixed nerve".

SPM – See: "selectively permeable".

starch – a polymer of glucose that is made by plant tissues as a way to store glucose. Starch is digested by amylase enzymes.

start codon – the initial codon in a mRNA strand that initializes translation. The start codon is AUG, which codes for methionine.

stem cell – a parent cell in bone marrow which gives rise to new blood cells. Unlike most cells in the body, stem cells are actively dividing.

steroid – a complex lipid. Steroids have a characteristic 4-ring structure. Cholesterol and some hormones (sex hormones) are steroids. See also: "cholesterol".

stimulus – a change in an environment. If a stimulus is strong enough (i.e., surpasses a threshold) it will have an impact on the organism (initiate and impulse).

stomach – a significant organ in the digestive system lying between the cardiac and pyloric sphincters. The stomach has three layers of muscles in its walls giving it the ability to churn food. The stomach has three major functions: (1) Storage of food material that has been swallowed. Typically the stomach can stretch to hold a volume of two or more liters. (2) Physical digestion, which is a direct result of the churning action. (3) Chemical digestion of protein, which is catalyzed by pepsin, a component of gastric juice.

stretch receptor – a type of nerve receptor that is sensitive to stretch. Stretch receptors are located on the surface of alveoli. They trigger impulses to the medulla oblongata that cause it to stop sending impulses to the diaphragm and intercostals muscles triggering exhalation.

subclavian artery/vein – blood vessels that travel under the clavicle (collar bone). The subclavian arteries are branches of the aorta that take blood to the body walls and shoulder areas. The brachial artery is a branch of the subclavian artery. The subclavian vein returns blood from these areas to the superior vena cava, which conducts the blood back to the right atrium.

substitution – one of three types of gene mutations that can affect the accurate replication of DNA. When a substitution occurs, the wrong nucleotide is incorporated into the new strand. Ultimately, this can affect a protein that is produced by the gene containing the substituted nucleotide, particularly if the location of the substitution is in the first or second position of the codon of the mRNA strand made. See also: "addition", "deletion", and "degenerative".

substrate – a reactant in an enzymatic reaction. Substrates are believed to form "activated" complexes with enzymes that facilitate the reaction. See also: "activated enzyme-substrate complex".

substrate concentration – the relative amount of substrate molecules in the solution where an enzymatic reaction occurs. Substrate concentration has an impact on the rate of product formation. Obviously, as substrate concentration increases, so too, can the rate of product formation. At some elevated level, the rate of reaction will no longer increase, as the enzyme concentration becomes the limiting factor.

sucrase – the disaccharidase that digests sucrose. Sucrase is a component of intestinal juice. See also: "intestinal juice", "disaccharide", and "disaccharidase".

sugar-phosphate backbone – refers to the backbone strands of nucleic acids. There are two strands in DNA and a single strand in RNA. Both are composed of alternating sugars and phosphate groups. In both cases, the nitrogenous bases are attached to the sugars.

surface area – the amount of surface a structure has. See: "SA/V ratio".

surface tension – the nature of a liquid to appear to have a "skin" on its surface. Water has high surface tension brought about by the cohesiveness of the molecules.

swallowing – a reflexive action that initiates the movement of a ball of food (bolus) down the esophagus. Swallowing is controlled by the medulla oblongata. The epiglottis covers the top of the trachea to facilitate swallowing.

symbiosis – a mutualistic relationship. *E. coli* exists in a symbiotic relationship in the large intestines of some mammals. The feces provide nourishment for the bacteria, and the mammals benefit from the bacterial metabolism.

sympathetic division – a subset of the autonomic nervous system (as opposed to the parasympathetic division). The sympathetic nerves consist of motor neurons that release noradrenaline to affect the effector. Noradrenaline will increase the heart and breathing rates yet decrease digestive activity. The sympathetic nerves leave the CNS from the thoracic and lumbar regions of the spinal cord. The sympathetic effector system requires two motor neurons to reach the effectors, meaning there are sympathetic ganglia (nerve cell bodies) outside of the CNS. These ganglia are located near the CNS.

synapse – the junction of a neuron and the next cell. See also: "synaptic gap".

synaptic ending – the ending of an axon at a synaptic gap. See also: "synaptic gap".

synaptic gap / cleft – the space between an axon and the next cell, be it another neuron, or an effector. This is the space that must be bridged by neurotransmitters for the continuation of the impulse. See also: "synaptic transmission", "neurotransmitter".

synaptic transmission – the chemical transmission of a nerve impulse across a synaptic gap. When a nerve impulse reaches the end of an axon, calcium ions, which are present in the synaptic gap, are moved into the axon terminus. These ions cause cytoskeletal proteins to shorten, drawing neurotransmitter-filled secretory vesicles to the presynaptic membrane. This is followed by the exocytosis and diffusion of the neurotransmitters. For successful transmission, a permeability change must occur to the postsynaptic membrane. This change occurs when receptor sites on the post-synaptic membrane receive the neurotransmitter.

synaptic vesicle – the secretory vesicle that delivers the neurotransmitters to the presynaptic membrane.

systemic circuit / circulation – the part of the circulatory system that delivers oxygenated blood to the body cells. It starts at the left ventricle and ends at the right atrium.

systole – See "systolic pressure".

systolic pressure (systole) – the force of the blood outwards on the arteries when the ventricles are contracting, as opposed to diastolic, the pressure between contractions. Normally, high blood pressure is in the range of 120 mmHg. See also: "diastolic pressure", "blood pressure".

target organ – with reference to hormones. A target organ is the organ (or tissue) affected by a particular hormone. For example, insulin affects the liver, its target organ to increase its uptake of glucose and conversion glucose into glycogen for storage. Other examples of target organs are the gonads that are affected by gonadotropic hormones.

temperature regulator – a key function of water in organisms. Because of its high specific heat capacity, water can withstand temperature changes even though the environmental temperature changes. This is of great assistance to organisms in terms of body temperature.

template – with reference to DNA. DNA acts as a template (or pattern) for the formation of mRNA during the process of transcription.

termination – ending. Termination refers to the final step of translation when a protein chain is ended. Termination involves the use of one of three terminator codons, which correspond to tRNA molecules that lack amino acids. When they are moved into position, the growing polypeptide chain can't bond to anything, therefore it is terminated. See also: "terminator codon".

terminator (stop) codon – one of three codons that does not code for an amino acid. These are used to stop the addition of amino acids into the sequence, thereby ending the protein. See also: "termination".

terminus – end. An amino acid has an amino terminus and a carboxyl terminus. A polypeptide has a sugar terminus and a phosphate terminus The word "terminus" is also used to describe the synaptic end of an axon.

tertiary structure – a level of protein structure that is apparent in many proteins. The bending and folding of the secondary structure into a three-dimensional shape characterize the tertiary structure of a protein. Enzymes are an example of a protein that has tertiary structure. These shapes are held together by hydrogen, ionic and covalent bonds that form between different parts of the molecule. The hydrogen bonds are the most sensitive to pH and temperature changes. Should they break, a protein with tertiary can denature.

testes – male gonads. There are two key tissues in the testes. The seminiferous tubules in which spermatogenesis occurs under the influence of FSH and the interstitial cells that produce testosterone under the influence of LH. LH and FSH are gonadotrophic hormones.

testosterone – steroid hormone that is produced by the interstitial cells of the tests and in small amonts by the adrenal cortex. In males, testosterone produces and maintains secondary sex characteristics (ancillary hair growth, enlargement of the larynx skeletal and muscle growth, maturation of the secondary sex characteristics, and etc.). In females, low levels of testosterone (from the adrenal cortex) contribute to muscle development and (what can be described as) a more masculine, aggressive nature.

thalamus – a region of the middle brain that is nicknamed the "sorting center". All impulses traveling to the cerebral cortex pass through the thalamus, which directs them to the correct association area.

theory – a hypothesis that has been supported by repeated testing.

thoracic – referring to chest. See also: "thoracic cavity".

thoracic cavity – chest cavity. The thoracic cavity contains the lungs, and heart. Its lower limit is the diaphragm.

threshold level – See also: "threshold value".

threshold value – the amount of a substance or stimulus required to cause an effect. Here are two examples: (1) In terms of nerve impulses, a stimulus must surpass a minimum threshold before it can open the sodium gates and cause an impulse. (2) Blood also has a threshold level of glucose. Amounts above the threshold will result in the excretion of glucose because the reabsorption process of glucose by the proximal convoluted tubule will not allow it to surpass that threshold.

thrombin – a soluble blood protein (enzyme) that is formed from the substrate prothrombin and catalyzed by thromboplastin during the blood clotting process. Calcium ions are required for this reaction. Thrombin catalyzes the conversion of fibrinogen to fibrin, which ultimately forms a blood clot.

thrombocyte – See: "platelet".

thromboplastin – an enzyme that is released by platelet when they are damaged. Thromboplastin catalyzes the conversion of prothrombin to thrombin, another enzyme. See also: "platelet", "thrombin", and "blood clot".

thymine – a nitrogenous base found in nucleic acids. Thymine is one of three bases called pyrimidines. The others are Cytosine (C) and Uracil (U). Uracil is found only in RNA and Cytosine is found in DNA and RNA, but Thymine (T) can only be found as a component of nucleotides in DNA. Thymine is complementary to Adenine, a purine base.

thyroid gland – an endocrine gland located in the neck region. The thyroid gland produces and secretes thyroxin. This hormone requires large quantities of iodine, which is accumulated from blood by the thyroid gland via active transport. See also: "thyroxin".

thyroxin – a hormone that accelerates cellular metabolism (the utilization of oxygen and the production of carbon dioxide). People with low thyroxin levels are often overweight have low energy levels. See also: "thyroid gland".

tissue – a group of the same kind of cells with a common function. There are many examples of tissues, such as: epithelial tissue that lines the villi of the ileum - specialized for absorption; blood, which is a complex tissue that is specialized to fight infection, transport oxygen, nutrients, and wastes, form blood clots, an help regulate the temperature of the body.

tissue fluid – (AKA extracellular fluid) is the fluid that is between cells within a tissue. Oxygen and nutrients move (through diffusion and facilitated transport, respectively) from this fluid into cells. Cellular wastes (ammonia and carbon dioxide) move (by diffusion) out of the cells into the tissue fluid to be taken away by blood. See also "capillary tissue fluid exchange".

tonicity – refers to concentration. See also: "hypertonic", "isotonic", and "hypotonic".

toxic – harmful. Some substances are toxic to cells. A good example is ammonia, which is a product of the deamination of amino acids. Ammonia is swept away from cells by the circulatory system. It gets converted to urea by hepatocytes (liver cells), which is less toxic, and finally removed from the blood by the kidneys during urine formation. At a cellular level, the SER detoxifies the cytoplasm. See also: "SER".

trachea – windpipe. The trachea is the cartilaginous ringed tube that conducts air up and down the neck. The tube itself is soft-walled, but the rings of C- cartilage hold it open, preventing it from collapse. The top of the trachea is specialized into a larynx (voice box) and has a ventral flap of tissue (epiglottis) that covers it, preventing the entry of food during swallowing. At the bottom, the trachea divides into two bronchi.

trans conformation – (as opposed to cis-conformation) the shape of an organic molecule that has a double bond where the longest parts of the molecule are on opposite sides of the double bond. Trans fats are unsaturated fats that are hard to digest.

transcription – the first of two processes that synthesize protein. During transcription, a section of DNA corresponding to a gene unzips to expose its bases. RNA nucleotides bond in a complementary fashion to the exposed bases. They then bond together to form a strand of RNA that is complementary to the gene. The RNA, called mRNA leaves the nucleus to go to a ribosome for translation, the second process in the production of protein. See also: "translation".

transfer RNA – an oddly-shaped molecule of RNA that has an anticodon on one end and an amino acid on the other. There are 64 different tRNA molecules (as there are 64 different codons). tRNA transports the amino acids to the ribosomes for the process of translation. See also: "translation".

transition vesicle - a vesicle produced by ER and containing "raw" products (either protein or lipid). A transition vesicle takes its contents to the Golgi apparatus where they are further prepared for secretion or incorporation into a lysosome, as in the case of some hydrolytic enzymes. See also: "Golgi apparatus".

translation – the second of two processes that synthesize protein. Translation starts with initiation, where the start codon (AUG) is recognized by two ribosomal subunits, which come together to form a complete ribosome. The ribosome utilizes enzymes to bring in tRNA molecules with complementary anticodons to the codons. The first tRNA molecule bears the anticodon UAC and the amino acid methionine. The ribosome "reads" each codon in the sequence and organizes the tRNA molecules into position. Once in position, the growing amino acid sequence is bonded by a peptide bond onto the newly arrived amino acid, thus the protein is elongated. The protein chain is ended, or terminated, when the final codon matches with an anticodon on a tRNA molecule that lacks an amino acid. Because of this, the polypeptide chain is unable to bond to anything, and breaks free as a completed protein. The used tRNA molecules return to the cytoplasm where they can become associated with another amino acid and used again. See also: "transcription".

transmembrane – refers to a protein that extends through a membrane. Transport proteins are typically transmembrane proteins, where receptors are not.

transmembrane protein – See "integral protein".

transparent – able to transmit light without (much) interference. Water is considered to be transparent, though it does cause the refraction of light. This transparency of water is valuable as it allows light to penetrate into the oceans for photosynthesis, eyes can be kept moist and animals can still see, etc.

triglyceride – a neutral fat consisting of glycerol combined with three fatty acids.

tripeptide – the sequence of three amino acids held together by peptide bonds.

triplet – a group of three. Codons and anticodons are triplets of nucleotides.

trisomy-21 – a genetic disorder (Downs Syndrome) caused by the incorrect separation of chromosome pairs #21 during gamete productions such that the zygote has three chromosome #21's.

trypsin – a protease enzyme component of pancreatic juice. Trypsin is produced as trypsinogen, a precursor. It is activated by the enzyme enterokinase before it enters the duodenum where it functions to digest protein (polypeptides).

trypsinogen – See: "trypsin".

tubular excretion – excretion process that occurs at the distal convoluted tubules in the kidneys. The blood in the peritubular capillaries contains some molecules that are no longer useful (though they have contributed to its hypertonic nature for the reabsorption of water). These join the filtrate at the distal convoluted tubules. This is how the body rids itself of penicillin and histamines. The pH of blood is also regulated (raised) in this way by the excretion of hydrogen ions. See also: "tubular secretion".

tubular secretion – similar to tubular excretion. However, tubular secretion moves substances back into blood. Such is the case for bicarbonate ions, which would raise the pH of blood, Na^{1+} (under the influence of aldosterone) etc. See also "tubular excretion".

tubulin – a protein that forms the structures of microtubules, part of the cytoskeleton of cells. Cilia and flagella are made out of microtubules. See also: "cilia", "flagella".

turgidity – See: "turgor pressure".

turgor – See: "turgor pressure".

turgor pressure – the pressure exerted by the cytoplasm against the cell membrane. In animal cells turgor pressure (turgidity) causes cells to swell. In plant cells, the effect is the same, however, the cell wall prevent noticeable swelling. See also: "deplasmolysis".

umbilical arteries/vein – the vascular components of the umbilical cord. See also: "umbilical cord".

umbilical cord – the blood vessels and associated tissue that connects a fetus (and embryo) to the placenta. An umbilical cord contains three blood vessels. The umbilical vein conducts blood with oxygen and nutrients into the fetus and anteriorly through the venous duct (through the liver) and joins with the inferior vena cava and into the fetal circulatory system. The two umbilical veins, which are branches of fetal iliac arteries carry blood from the fetus to the placenta for the removal of toxins like CO_2 and NH_3.

unit molecule – See: "monomer".

unsaturated fatty acid – a long-chain carboxylic acid (fatty acid) that has one or more double bonds in its carbon chain, and is therefore not saturated with hydrogen. See also: "saturated fatty acid".

upswing – (as opposed to downswing) the portion of a graph showing an action potential that characterizes the effect of depolarization. During depolarization, the Na^{1+} gates are open and Na^{1+} floods to the inside changing the relative charge of the axoplasm from –65 mV to +40 mV. See also: "downswing", "depolarization".

uracil – a nitrogenous base found in nucleic acids. Uracil is one of three bases called pyrimidines. The others are Cytosine (C) and Thymine (T.) Thymine is found only in DNA and Cytosine is found in DNA and RNA, but Uracil (U) can only be found as a component of nucleotides in RNA. Uracil is complementary to Adenine, a purine base.

urea – a toxic nitrogenous waste made from ammonia (NH_3)by the liver. Ammonia is made from the deamination of amino acids. It is very toxic. The liver cells convert it to urea, $CO(NH_2)_2$, which is less toxic. It is finally excreted in urine. See also: "ammonia", "toxin", and "urine".

ureter – the muscular tube that uses peristalsis to conduct urine from the urinary pelvis in each kidney to the common urinary bladder where it is temporarily stored.

urethra – the muscular tube that conducts urine from the urinary bladder to the outside during excretion. The movement of urine into and out of the urethra is controlled by two sets of sphincter muscles.

urethral opening – the terminus of the urethra. See: "urethra".

urinary bladder – the muscular organ that stores urine until it is excreted. Urine is conducted to the bladder through the ureters and it leaves through the urethra.

urine – the fluid waste of the body that is excreted through the urethra. Urine gets its name from urea, the toxic waste that it contains.

uterine cycle – the repeating changes that occur in a uterus. The uterine cycle has three phases: menses, (lasting approximately five days), proliferation, (lasting until the glands become secretory - approximately nine days), and the secretory phase, (lasting until menstruation starts again.) The changes in the uterus are caused by changes in the levels of estrogen and progesterone. See also: "menstruation", "proliferation", and "secretory phase".

uterus – (AKA "womb") – a muscular organ designed for implantation and fetal development. The uterus has a thick-walled myometrium (muscle layer) and an inner vascular layer. The changes that occur to the endometrium are known as the uterine cycle. See also: "implantation", "uterine cycle".

vacuole – the organelle characteristic of plant cells that is used for the storage of excess water and other substances. In animal cells food vacuoles exist following phagocytosis. See also: "phagocytosis".

vagina – the female organ of copulation, also known as the birth canal.

validity – a measure of the usefulness of the results of scientific investigation. If a scientific investigation did not fairly measure what it was intended to, then the results would be considered "invalid".

valve – a structure that controls the movement of fluids along a tube. Sphincter muscles are often called valves. Veins, the heart and the lymphatic system contain valves to prevent the backflow of blood.

vascular – refers to blood vessels. The alveoli are highly vascularized, meaning they have a rich supply of blood vessels.

vegetative – as opposed to "active". The parasympathetic nervous system promotes vegetative body control, which includes decreased heart rate, breathing rate, increased digestive action etc. – all the things associated with resting/sleeping.

vein – blood vessel that returns blood towards the heart. Veins have valves to prevent the backflow of blood.

vena cavae – the major veins that conduct blood into the right atrium. The anterior (superior) vena cava drains blood from the anterior regions of the body (above the heart), while the posterior (inferior) vena cava drains blood from the lower regions of the body (below the heart).

venous duct – the tube-like portion of the umbilical vein that courses through the liver, thus blood can bypass the functions of the liver. This is essential so it can get to the heart and be pumped around the fetal body. After birth, the venous duct atrophies.

ventral root – a nerve extension (process) extending out from the ventro-lateral part of the spinal cord. Motor neurons leave the CNS through ventral root.

ventricle – a chamber. In the heart, the ventricles are the pumping chambers. The left ventricle marks the beginning of the systemic circuit, where the right ventricle marks the beginning of the pulmonary circuit. In the brain, a ventricle is a chamber that is filled with cerebral spinal fluid.

venule – a small vein.

vertebrae – the segmental bones that make up the vertebral column. The spinal cord passes through the middle of the vertebrae and is thus protected from impact.

vesicle – an organelle that contains materials for transport in a cell. Vesicles are made by the ER, Golgi apparatus and by the cell membrane. See also: "secretory pathway", "endocytosis".

vestigial structure – a structure that is not functional or useful. The appendix is an example of a vestigial structure.

villi – the finger-like projections along the inner wall of the ileum. These contain blood capillaries and a lacteal and function for the absorption of the products of digestion.

virus – non-living particles consisting of protein and nucleic acid with the ability to invade cells and control their metabolism. Viruses are considered to be foreign antigens and are attacked by white blood cells.

viscosity – "thickness". Fluids have different viscosities. The rate of diffusion through the fluids is impacted by the viscosity of the fluid – the lower the viscosity, the faster the diffusion.

vitamin – a chemical substance that assists enzyme function. There are different types of vitamins, (A, B, C etc.) Vitamins A, D, E, and K are fat-soluble and the rest are water-soluble. NAD is a complex that contains niacin, one of the B vitamins. It serves as a hydrogen carrier. Vitamins are absorbed along with digested food nutrients in the intestines. *E. coli* in the large intestine actually makes small amounts of some vitamins, which are absorbed along with water from the forming feces.

vocal cords – two fleshy tendons located in the larynx. The tautness (tightness) of the vocal cords causes them to vibrate at different frequencies, which result in the different pitches of sounds used in vocalization. See also: "larynx".

wax – a lipid that is a combination of fatty acids bonded to alcohols longer than glycerol.

white blood cell (AKA: leukocyte) – the type of blood cell that combats infection. There are several types of white blood cells, all produced from bone marrow, but specializing in different ways providing the body with a variety of mechanisms to combat infection. These mechanisms include phagocytosis, releasing antibodies to cause agglutination, releasing enzymes that destroy foreign cell membranes and etc. Generally speaking, white blood cells are about twice the size of red blood cells, are not a abundant and often have cytoplasmic granules and a complex shaped nucleus. See also: "agglutinate", "antibody".

white matter – This is the outer, lighter-colored portion of the CNS. In the spinal regions, the white matter is one the outside, whereas in the brain, the white matter is on the inside. The lighter color of the white matter is a result of the abundance of myelin. The white matter contains nerve tracts.

womb – See: "uterus".

X-ray – a type of radiation that is known to cause gene mutations when DNA is exposed to high doses of it.